Praise for *Little Steps, Big Faith*

"*Little Steps, Big Faith* beautifully weaves together science, faith, and parenting. A welcome handbook for parents and pastors alike, Rundman's book utilizes her expertise in developmental psychology to give scientific rationale to the importance of steeping infants and toddlers in the language and practice of the Christian faith. Accessible, biblical, and practical, *Little Steps, Big Faith* is the perfect gift for new parents, baptismal sponsors, and anyone wishing to pass on the faith to the next generation."

—Justin Lind-Ayres, pastor and author of
Is That Poop on My Arm? Parenting While Christian

"With *Little Steps, Big Faith*, Dawn Rundman expertly helps parents fuse their Christian faith with the important milestones of child development. The real-life examples are utterly charming and help prepare the reader for the many surprises of parenting."

—Justin Roberts, Grammy Award–nominated
songwriter for kids and families, and author of
The Smallest Girl in the Smallest Grade

"This book is a good fit for parents and caregivers who take seriously the opportunity to invest their faith into the lives of their young ones. Real-life stories and 'Little Steps' cement the concepts so that caregivers can easily see the relationship between their activity and forming faith in their child."

—Rev. Dr. Phyllis B. Milton, Hampton, VA

Little Steps
Big Faith

How the Science of Early Childhood Development
Can Help You Grow Your Child's Faith

Dawn Rundman

beaming
books
MINNEAPOLIS

For my parents Dean and Elaine, whose faith, hope, and love
nurtured me (and my brain)

For Jonathan, my beloved partner in life together

For my children Paavo and Svea, whose beautiful, faithful selves
unfold before my very eyes

Published in 2018 by Beaming Books, an imprint of 1517 Media. All rights reserved. No part of this book may be reproduced without the written permission of the publisher. Email copyright@1517.media. Printed in the USA.

Scripture quotations are from the New Revised Standard Version Bible, copyright © 1989 National Council of the Churches of Christ in the United States of America. Used by permission. All rights reserved worldwide.

ISBN: 978-1-5064-4685-1
Ebook ISBN: 978-1-5064-4830-5

Library of Congress Control Number: 2018951000

Beaming Books
510 Marquette Avenue
Minneapolis, MN 55402
Beamingbooks.com

Contents

1

Child Development + Faith: An Introduction

I began my career in early childhood development as a psychology professor at a Lutheran college well known for its teacher education programs. A student in my Infant Development course one fall semester has always stuck with me. As a business major among education majors, and the only man enrolled in my class, Brad stood out most because of his unique role—he was a new father who chose this elective class to learn more about his baby daughter.

Besides his stories of his newborn, a memorable thing about Brad was his devotion to the University of Tennessee football team. Most weeks that fall he wore a bright orange UT letter jacket and baseball cap, and sometimes he had a team logo t-shirt or jersey on underneath. Since it was football season, he mentioned that on game days, he'd watch the UT game while holding his baby girl. He described how she'd get worked up during big plays, her arms and legs wiggling with delight because she sensed her dad's excitement. When his wife brought their daughter to class so everyone could finally meet her, she was wearing a UT outfit.

One week, Brad and another student arrived several min-utes early. As they waited for class to begin, Brad shared that his family members were gathering that Sunday at the university's chapel, where the campus pastor would baptize his daughter. The other student asked why they chose the university's chapel for the baptism instead of their home congregation. By now I was listening intently, although I tried to look busy preparing for the upcoming class.

Brad explained that he and his wife weren't members of any church at that time. Then he said something I vividly remember over two decades later: "We thought we'd get her baptized now, and then when she grows up, she can decide."

This new father, clearly devoted to his child, appeared to be doing everything within his power to raise a UT foot-ball fan. But when it came to laying the foundations for his daughter's faith formation, he was asserting little or no influence. Instead of claiming his position as one of the two people who could most significantly shape his daughter's earliest experiences with God, faith, and the church, Brad adopted a wait-and-see perspective. He assumed that his daughter would somehow take up her own explorations of faith without any initiative, support, or enthusiasm from her parents.

Why This Book?

I believe Brad held a flawed belief about the guiding forces that could shape his daughter's faith. In class, he was taking many positive steps toward being an involved, loving father by learning about her development in ways that helped him parent more effectively. But Brad needed ideas and affirmation to help him connect these early-childhood insights to ways he could nurture the beginnings of her faith, hopefully with the same enthusiasm and commitment he showed on game days.

I've spent close to three decades teaching parents, church leaders, and educators how to make connections between early childhood development and faith formation. I'm convinced that parents of young children can benefit from learning about these intersections, allowing them to confidently embrace parenting practices that support their children's budding understandings of God, faith, and the world around them. This book's purpose is to affirm two points about you, your child, and your child's faith:

- The first three years of life are a remarkable time of development for children, with many critical windows when early experiences can make lasting impressions.

- Shaping a lasting foundation of faith in your child doesn't require you to make sweeping changes to your family life—only to open up to seeing how everyday family life can be viewed through a faith lens and then adjusted to incorporate faith moments into these everyday times.

The First Three Years

Experts and researchers in the field of early childhood development have studied children in laboratory settings for over a century. For decades, these experts shared their findings with each other in peer-reviewed journals, with some insights making their way into a few best-selling parenting books. But today, the ubiquity of technology gives parents unprecedented access to information, advice, and support relating to their children's development. As an expectant parent, you can track prenatal development daily or connect regularly with those involved in your adoption process, search online for any parenting topic imaginable, and find the bloggers who offer their own answers to your most burning parenting questions.

Never before have parents had access to so much information about development during the first three years of their children's lives. The media, technology, toys, and classes now available that align with these research findings are probably mind-boggling to grandparents, most of whom didn't have these resources available to them when they were parents years ago. (Ask any first-time grandparent about their inaugural visit to a baby superstore to get them started on how times have changed.)

But this information about child development rarely, if ever, addresses the spiritual development of children. While parents have options galore to learn about growth trajectories, tips for first foods, and suggestions for getting a baby to sleep through the night, few resources help parents form a perspective on how faith formation can align with the thrilling developments of the first three years. During this time of your child's life, you have an amazing window of opportunity. (This is where my student Brad could have used some encouragement and support.) *You* get to decide. *You* get to choose. *You* get to create many of the first experiences for your young child to encounter God's presence.

This book is designed to help you gain confidence and assurance in this faith-formation role. You'll learn why this role is such a crucial one for the sake of your child in the first three years of life. You may be worried that you'll have to learn new techniques, start new routines, and try really uncomfortable things to be successful. This book will help you see that *you are already doing* many of the things that can shape their faith. It's just that you haven't deliberately thought about how the things you're already doing could influence your child in this way.

What's in This Book

In these chapters, you'll read about what child development researchers have learned regarding this critical window of the infant and toddler years. Advances in research technology, paired with the persistence, hard work, and ingenuity of those working in research labs, have provided new and exciting ways of understanding and describing human development in the first three years. Hundreds of books focus on these amazing findings to help mothers and fathers apply them to their parenting practices. But this book will help you see how these results can be viewed in new ways when you apply a faith lens.

I begin each chapter with a story to bring its topic into focus. Then I summarize research from important child development studies to identify key milestones and advances. I'll help you view these findings through a faith lens to provide insights into how you can infuse faith into daily life with your young child. Each chapter ends with some "Little Steps" to try. You don't need to try the whole list; just pick one or two that fit into your busy schedule.

In chapter 2 on brain development, I'll use language typical of gardening to help you visualize how experiences form and strengthen pathways in your child's brain. I'll also introduce you to some natural and easy ways to begin to structure faith experiences with your child.

Chapter 3 summarizes research on the vital importance of attachment relationships to demonstrate the essential nature of the parent-child bond. I'll introduce you to a way of viewing attachment through a faith lens so you can marvel at how the parent-child relationship is the first way little ones learn about their relationship with God.

In chapters 4, 5, and 6, I'll share some amazing research findings on language, literacy, and music during the first three years of life. Talking, reading, and making music together can be everyday activities with your little one. Adding the faith perspective opens up possibilities for children to begin experiencing God's presence in their lives, hearing stories of God's people, and learning songs that praise God.

Chapters 7 and 8 emphasize the many opportunities for faith moments during our physical care of our little ones. Touch and movement can be included within new rituals of blessing and comfort for young children.

Chapter 9 provides an in-depth look at how the church is a rich context for child development by highlighting all the ways that church communities support attachment, language, literacy, music, physical contact, and rituals.

And in chapter 10, I'll wrap things up by giving you some additional ideas for little steps of faith. By this book's conclusion, I hope you will feel affirmed in your role as a caregiver who can shape experiences that help form a strong foundation for your child's faith to grow.

Here are a few notes to keep in mind as you read:
- Throughout this book, I use the word *parent* to describe those primary caregivers who are committed to the daily

care and lifelong relationship with a child. *Parent* can refer to biological, adoptive, or foster parents, along with others who are legal guardians. The ideas in this book can also be used by extended family members such as grandparents, aunts, and uncles, especially if they provide regular care during a child's first years of life.

- For the sake of brevity and simplicity, the pronouns *they* and *them* are often used to refer to an individual child, rather than him/her or she/he.

- Most of the developmental research presented pertains to children experiencing typical development. I have tried to create faith-based content that will also be relevant for parents whose children have developmental delays, disabilities, illness, or injuries that significantly affect development.

As you read this book, I invite you to be open to deepening your awareness of ways to weave faith into daily life with your young child. Be ready for new insights that may lead to different parenting choices. You've taken a great step by reading a book that connects child development with faith formation.

Are you ready? Let's get started!

2

Brain Development: Experiences Shape Pathways

When I was in graduate school, I met Andrew, a friend of a friend who was both the father of a toddler and an avid skydiver. (As if parenting a toddler isn't thrilling enough!) He was interested in my early-childhood studies, and we soon struck up a conversation about the amazing speed at which young children learn new skills. He shared an insight from the time his daughter began to walk because it paralleled a skydiving experience of his.

When he first joined a group of skydivers and began to dive in formation, the process of moving through the air to the correct position required intense concentration. He described the need to focus his attention on specific parts of his body, moving them just so, shifting his weight, and adjusting his actions second by second so he could make it to the correct spot in time to take his place amid the others. However, he said, after several jumps with this group, that movement became much easier. "It was almost as if I could look at the place I was supposed to be and move there automatically."

Then he described a similar transformation in his daughter as she learned to walk. When she started to walk, he could see her look to where she wanted to go. "If she wanted to walk to the other side of the coffee table," he told me, "she would seem completely focused on every part of her body, shifting her weight and making adjustments so she could clear the edges." But after many steps (and falls), she gained the smooth, fluid movements of a confident toddler. Once she was walking, it was almost as if she could look at the other side of the coffee table and move there automatically, just as her father had done for his skydiving configurations.

A child's first steps may appear clumsy, but they mark a striking advance for one who has been working for months to develop the balance, coordination, muscle strength, and persistence to finally stand on their own two feet. This rapid acquisition of new motor skills is just one way that young children show the power of brain pathways forming and deepening with repeated experiences—just as a skydiving adult can make new connections with every jump. Fortunately, you don't need to be a skydiver to notice the concentration of your little one as they learn new motor skills, whether they're holding a piece of food, sitting up without support, or taking their first steps.

Learning some basics about your child's brain development can help you more deeply understand *why* these changes are happening. You will also see how the environment and experiences you create for your child at home and on the go can support healthy brain development. And once you step more knowingly into this role of supporting your child's brain development, you can take little steps to help your child literally wire the building blocks of faith into their brain.

Brain 101

Your baby was born with about 85 billion brain cells, called neurons. Talk about a grand entrance! Neurons duplicate exponentially during prenatal development. After birth, your child's brain development continues at a similarly fast pace, but in a different way.

To understand how your child's brain develops in the first three years, it helps to know a little bit about the structure of their brain cells. (Ready for a quick crash course in Neuroscience 101? Let's go!) We'll start with **dendrites**, cellular structures that look like bare tree branches in winter. Dendrites receive messages from surrounding brain cells. Once a dendrite is activated by a neighboring neuron, a nerve impulse travels along the cell membrane and reaches another main structure, called the **axon**.

When the impulse reaches the end of the axon, it triggers action in the axon terminal, where neurotransmitters such as dopamine, serotonin, and norepinephrine are stored. These neurochemicals are released into the space, called a **synapse**, between the axon and the surrounding dendrites. If these neurotransmitters stay in the synapse long enough, they stimulate surrounding dendrites to produce a neural impulse that is strong enough to start the whole process again.

Getting Wired

After birth, your child's neurological development is not marked by growing more brain cells. Instead, the neurons your little one already has become more efficient in communicating with each other. Experts have estimated that in the first few years of life, more than a million new neural connections are formed every second!

As neurons communicate with neighboring cells, eventually they form pathways that connect into networks. Rapid brain growth in the first years of life happens because of these new connections. Cells that communicate with each other more frequently will build stronger, richer pathways. Pathways that form but are not regularly activated can weaken or disappear. This process of pathway formation and loss is summed up in the phrases "Neurons that fire together wire together" and "Use them or lose them."

Brains Are Botanical . . .

Neural pathways are shaped by two processes with names that may remind you more of gardening than gray matter: dendritic arborization and synaptic pruning. **Dendritic arborization** occurs when a cell grows additional dendrites in response to repeated experiences. (Think of the difference between a young sapling and a mature tree in terms of the number of individual branches and the overall spread.) These cellular connections show up in a child's everyday actions, like when your baby's clumsy swipes for a cereal puff eventually transform into their fingers selecting food with surgical precision. If you have a toddler who's a climber, you've seen how quickly coordination and speed can develop after those clumsy first steps.

These repeated experiences, whether grabbing, walking, or climbing, lead to denser dendrite networks. But just as a tree needs to be thinned out in order to grow stronger and healthier, networks need trimming too. A process called **synaptic pruning** is a necessary part of brain development. If all dendritic connections kept expanding, the brain would become slower and less efficient. The process of pruning occurs to connections that are not used as frequently so the brain can transmit impulses efficiently. (That's the "lose them" part of "Use them or lose them" at work.)

... And Plastic

Most babies crawl, walk, smile, talk, and play on a predictable schedule because of how pathways form as part of typical development. But other networks form in your child's brain because of the unique experiences they have each day. As a parent, you play a key role in helping your child form neural pathways that support healthy development. For example, if your child hears music regularly, they will develop pathways that recognize melody, rhythm, and pitch. Talking to your child from the very first diaper change builds connections throughout their brain's language centers to recognize words and, eventually, say first words.

The first three to five years of your child's life are the time of greatest neuroplasticity, meaning the brain is most open and able to form new pathways in response to experiences. Your little one's plastic brain is formed and shaped by many factors—and you are one of the main influences! You are designing your child's experiences, from the way you wake them up in the morning to the bedtime routines you enjoy together. Your activity patterns will evolve as schedules change and your child grows, but the underlying experiences help their brain build positive pathways for healthy development during all their encounters with the people, places, and things of their day.

Little Steps

As a parent, you are incredibly influential in determining what your child's earliest experiences will be. You have opportunities *every single day* to help their brain form rich networks. Some of those experiences can support faith formation during your child's earliest years. Think about your daily routines with your little one. What do you experience together just about every day? Here are some possibilities:

- Waking up your child

- Changing your child's diaper

- Dressing your child in clothing for the day

- Dressing your child in pajamas for the night

- Feeding your child morning, noon, and night

- Bathing your child

- Washing your child's hands and face

- Comforting your child

- Reading to your child

- Playing with your child

- Singing to your child

- Dancing with your child

- Getting your child in and out of the car seat

In the following chapters, you'll learn ways to introduce faith practices into these everyday times with your child. Start by noticing your routine activities, then building ways for such everyday moments to support your child's little steps toward big faith.

3

Attachment Schemas: Your Love Shows God's Love

February, 1954: After saying goodbye to his wife, Mildred, and their baby daughter, Kathy, Bert left for a tour of duty in Korea, where he was deployed for sixteen months.

Throughout his absence, Mildred wrote a letter to Bert almost every day and walked to the post office with Kathy to mail it. Each night before bed they would look at a framed photo of Bert, and Mildred would tell Kathy, "This is your daddy." Then Kathy would kiss the picture. "Daddy is coming back," Mildred assured Kathy, even though she knew the dangers her husband faced overseas. For sixteen months they waited anxiously for his return.

Finally, the day arrived. After an eight-day journey by ship, Bert docked in Seattle and joyfully greeted his wife and daughter. It was a wonderful reunion and homecoming! By the time he completed all the procedural duties at the base, it was too late to make the 300-mile drive home, so they stayed in a hotel that first night.

Kathy was used to sleeping in a crib, but that night she had to sleep on a small cot without siderails. In the middle of the night, she rolled onto the floor. As she fell, Mildred and Bert describe what happened next.

"She immediately called out, 'Daddy!'" Mildred says.

"Over sixty years later," Bert recalls, "whenever I think about that event, I still get a bit choked up and am touched that she called for me."

Despite all her father's time away, Kathy knew who he was, trusted him, and called out to him for help. She had formed an understanding of her relationship with him because Mildred faithfully assured her how much he loved her. Kathy had learned that she could call out to her father and he would respond to her and take care of her. What love!

Attachment occurs when a child become emotionally connected to a loving caregiver, often a parent. A child's attachment to a caregiver is a foundational relationship that begins forming in the first weeks and months after birth. Young infants show utter dependence on their caregivers for meeting basic needs, like feeding them when they're hungry, keeping them warm or cool enough, and changing their diapers. Typically, a baby develops this relationship with those who meet her or his most immediate physical needs, especially if the caregiver also provides gentle touch and comfort, even when it's not a diaper change.

An infant's focus, understandably, is on having their needs for food, comfort, and safety met in the moment. But as babies grow into toddlers, they become cognitively capable of forming ideas about people and things that are not immediately present in their environment. These mental models show a deeper understanding of their parents' love and care: *I know my parent loves me, even if I can't see him/her right now.* From there, a child can form a mental model of knowing that God loves and cares for them too, even though they cannot see God.

You can help your child connect the love and care they receive from you to the love and care of God. Just as Mildred helped Kathy know her father and trust him to comfort her even though she could remember no past experiences with him as her caregiver, you can teach your child about God, their heavenly parent. Your assurance sends a message that God loves them, cares for them, and is always present in their lives—even when they can't see or touch God. Your loving relationship with your child is their first and most vibrant example of this divine love.

Basics of Attachment Theory and Research

Attachment theory, studied by human development researchers, gives parents a window for understanding these close relationships in deeper ways. If you've taken an Intro to Psych class, you may remember your textbook describing the work of pioneering researchers who demonstrated just how crucial attachment is. (Note: Attachment theory is a different concept than the philosophy of attachment parenting, which guides some parents to practices such as co-sleeping and babywearing.) Here are what a few of the researchers discovered about the importance of the relationship between young children and the adults who feed, shelter, and love them.

John Bowlby's Work: In the 1940s and 1950s, John Bowlby and colleagues studied several motherless children who stayed in orphanages during and after World War II. Bowlby discovered that the lack of a maternal relationship early in life had lasting effects on later development in these children. Bowlby theorized that when young children do not have a secure base from which to receive comfort and explore the world, they are at risk for many developmental challenges, including their ability to form close, healthy relationships later in childhood, adolescence, and adulthood.

Harry Harlow's Monkeys: Also in the 1950s, Harry Harlow designed studies that deprived young monkeys of any care from their monkey mothers. His team constructed replacement monkey-shaped figures of chicken wire and wood that held a bottle during feeding times. Half of these wire-and-wood "mothers" were covered with soft terry cloth; the others kept sharp edges intact. Harlow and his colleagues discovered that the monkeys preferred contact with cloth mothers over the wire-and-wood ones, even when the wire-and-wood ones were the structures that held bottles. Those monkeys with cloth mothers explored their environments more thoroughly and returned more readily to the cloth mothers for comfort and security compared to the monkeys with wire mothers. (A quick image search for "Harry Harlow's monkeys" will show you what these deprived monkeys looked like. It may also make you want to hug your child's stuffed-animal monkey.)

Mary Ainsworth's Strange Situation: Later researchers did not use such stark circumstances to illustrate the power of early relationships. Mary Ainsworth, one of John Bowlby's colleagues, created a laboratory task called the Strange Situation, still used in research labs today. In the Strange Situation, a toddler and parent enter a room with some toys. Over the course of twenty-one minutes, the child faces short time periods of being with their parent, with an adult stranger (often a graduate student working in the research lab), with both adults, and also being left alone in the room.

Based on video recordings, researchers observed child behaviors during separations and reunions with the parent, along with their reactions to being with the stranger and being alone. These observations led to four categories of early attachment relationships: secure, insecure-avoidant, insecure-ambivalent, and insecure-disorganized. These classifications describing parent-child attachment during the toddler years often remain consistent in later years.

Early Attachment Relationships Predict Later Outcomes

A child's attachment relationship quality has major implications later in life. To demonstrate just how important attachment is, many studies have followed children whose attachment relationships were initially assessed in the toddler years. Children who show a secure attachment relationship with a parent or other caregiver as toddlers are more likely to have positive relationships with other family members and peers. Additionally, these children show increased competency in several other areas, including empathy, altruism, and school readiness.

Toddlers with insecure attachments to a parental figure, however, are at risk for negative outcomes later in life, including defiant behaviors, poor peer relationships, and anxiety. More recent research on the neuroscience of attachment and the effects of trauma highlight ways that the brain is wired to form strong relationships in the first years of life. When a child has no warm, loving caregiver, or when the only available caregiver provides harsh, abusive care, the brain begins to wire in different ways.

This body of research demonstrates that your child's attachments to you and your spouse/partner or other adult caretakers set the stage for them to learn about themselves, about you, and about what they can expect from the world around them. Parents who actively seek ways to form loving, healthy relationships with their babies and toddlers are already on the right track by recognizing the importance of developing secure, loving relationships in the early years.

From Secure Attachment Springs Faith

So, what do attachment relationships have to do with helping your child develop a lasting faith? Plenty, because faith is rooted in relationship. As Christians, we believe in a loving God who is in relationship with us and who loves us first. And God sent Jesus, who entered into relationship with all kinds of people to live out the greatest commandment of loving God and loving our neighbor.

To look more closely at how God calls us to be in relationship, think about the metaphors we use to understand who God is. Many images for God involve powerful terms like *creator*, *judge*, and *king*. But the simplest and most relevant image to use with your young child is God as parent. A parent who unconditionally loves. A parent who is patient, gentle, and kind. A parent who offers healing and care. A parent who forgives. A parent who teaches children how to live in loving relationship with others. This list could go on and on! Reading it may make you feel that as your child's earthly parent, you can't ever measure up. However, it may be that your relationship with your child gives *you*—as well as your little one—the deepest glimpse (and assurance!) of God's love for us.

During your child's first year, you build the foundations of your attachment relationship. Your child is utterly dependent on you for everything. By providing food, comfort, and safety, you meet their needs again and again and again. You swaddle and rock. You feed and then burp to help comfort a bubbly tummy. You murmur messages of love while they fuss, before they sleep, and when they wake. You are deeply tuned in to this little one and know more about them than anyone else!

And while it may take a few months to start seeing specific responses in your child, soon your child begins developing neural pathways that recognize your face, your voice, even your smell! These repeated experiences with you help your child develop a deep sense of trust that their needs will be met, no matter what, by you and their other primary caregivers. Your deep love for your precious child fuels your responses to the intense demands of the first year—how else to explain what helps you get through thousands of diaper changes, hundreds of feedings, nights of disrupted sleep, and the massive rearrangement of family schedules and priorities?

For adoptive and foster parents, this foundation-building may begin later than the first year, but many of your parenting activities parallel those with infants. You must tune in to your child's toileting needs, their feeding schedules and food preferences, and their nap and nighttime sleep habits. You must also learn what you can about their previous relationships with caregivers and address the challenges that arise if they have experienced relationships characterized by mistrust. Laying a foundation for secure attachment may be challenging, but brain neuroplasticity can help! During the first five years of life, your child's brain is especially designed to reshape neural pathways based on experience, writing over negative ones and forming new positive pathways based on your responsive, loving care.

A young child's dependence on their parents can prompt parents to think about how we depend on God. God, the source of life, has created us. God has known you since you were knitted together in your mother's womb (Psalm 139:13-16). God knows your heart (Psalm 44:21) and provides daily bread (Psalm 145:15; Matthew 6:11). God is faithful and assures us that we will be loved and cared for, no matter what. Your child's reliance on you can mirror your trust in and reliance on God.

As your child enters the toddler years, what they need from you as a parent changes. As your attachment relationship evolves, you will move into some new roles. You are still providing for their basic needs, but their world is expanding. Speaking first words, taking first steps, and other developmental milestones pave the way to the independence that toddlers are known for! As your little one explores the world around them, they can venture farther into what's new, but they still need to know that you are there. Your role is now to be a secure base providing reassurance, encouragement, and comfort.

Your child may feel confident in exploring their environment, trying new things, and pushing limits. However, once something unfamiliar or ambiguous happens, your attachment relationship is activated in a different way. Think about what happens when a toddler's play is interrupted by a clap of thunder during a storm, or when you visit a friend your child has never met and they won't leave your side (or lap) despite your assurances. It's as if your child is saying, "I know it may look like I want to do everything myself, but I'm checking to make sure you're there if I need you."

Again, we can draw parallels between the expressions of attachment in young children and the ways we connect to God. When we encounter what's unfamiliar, ambiguous, or threatening, our faith leads us to seek God. We know that when we face uncertainty or danger, God will be there for us, drawing us closer. The Old Testament stories of people called by God tell about their times of insecurity, failings, and fear—and they are always met by a God who is steady and loving and present. The book of Psalms is filled with verses about God's presence during times of trial and lament. We know that as Jesus faced death on the cross, he called out to his heavenly Father with the words "Abba! Father!" (Mark 14:36). The apostle Paul repeats this plea of "Abba! Father!" in Romans 8:15 and Galatians 4:6, pointing to how we are beloved children of God.

Modeling a Divine Love

You've heard it before: your child is utterly dependent on you for survival. They need you to provide for them and protect them. When you offer loving care that is responsive to their needs, you're on your way to forming a secure attachment relationship. (Add "secure base" to your job title as parent!) This loving relationship will help your child thrive developmentally and will shape the relationships they'll have throughout their life.

As a parent, you are an agent of God's love in your child's life. Developing a secure attachment relationship with your young child gives them their first window into their relationship with their Creator. When your child experiences the ways that you love, care for, and protect them, they are also witnessing what God's love, care, and protection are like.

Forming this secure, loving relationship with your child gives them a foundation for knowing how they're loved by God. When your child knows they are deeply loved for who they are, you are opening for them a window into God's grace that is given freely without any of our actions earning it. Living out your faith as an expression of your relationship with God can help your child begin forming their own understanding of this relationship.

Little Steps

- In your prayers, ask God to bless your relationship with your child and to give you the wisdom, strength, kindness, and whatever other qualities you need in your daily parenting practices.

- Use parent language when you talk about God. Whether you begin a prayer with "Our Father" or speak the Bible verse about the mother hen who protects her chicks (Luke 13:34), you can help your child envision a loving God who is like a heavenly parent to them. Your child still has a few years ahead of thinking that you're perfect—savor that! Through this language, your child can begin establishing a lifelong understanding about God's everlasting love and care for them by seeing all the ways you parent them.

• Reflect on your attachment relationships with your own parents or other primary caregivers. Becoming a parent yourself puts these bonds in sharp focus. If you need to talk through parts of this relationship that were difficult, challenging, or traumatic in your past, give yourself permission to confide in your spouse, a close friend, a pastor, or a mental health professional.

• Look to Scripture for examples of parenting qualities you can focus on in your own parenting life. The fruit of the Spirit in Galatians is a good place to start. Aiming for all seven qualities at once will feel overwhelming, so center on one at a time to see how you can more deliberately show it in your relationship with your child. What does gentleness look like when you are bathing your child? How could patience be part of your car-seat buckling routine?

• Pair physical care of your child with your words. "I love you so much!" spoken while you give an embrace supports little ones in associating your loving touch with loving speech long before they understand the individual words.

• Affirm the roles that other loving adults or older kids may play as attachment figures. Day care providers, grandparents, close friends, older siblings, and others who see the child regularly are developing an attachment relationship with your child. Make sure you thank these other attachment figures. If they are Christians, let them know that faith language is welcome and encouraged.

4

Language: What We Say Matters

When my son Paavo was a baby, my husband, Jonathan, stayed home to care for him while I worked full time. During that first year, Jonathan's parents drove eight hours from their home several times to visit and help us. Thrilled to spend time with their first grandchild, they showered him with love, attention, and fun. Grandma Mary needed no encouragement from me to read to this little child. She would hold him and read to him from a stack of board books, her kind, singsong voice lilting and dipping as she told him story after story.

During one of their visits when Paavo was eight months old, Jonathan greeted me excitedly when I returned home from work. "You'll never believe what Paavo did today!" he said. "Let my mom show you."

We walked to where Mary was playing with her infant grandson. With a big smile, she grabbed a board book. (It was a classic, Goodnight Moon*.) She began to read* *its pages, pausing at the spread with the reddish-orange round balloon. "Paavo," she asked, "where is the balloon?"*

My eight-month-old son raised his little finger, moving it toward the page. As he pointed unmistakably to the round shape in the picture, I felt utter astonishment. Still four months away from speaking his first word, he was demonstrating a remarkable language milestone I didn't even see coming. He could recognize a spoken word and then point to a representation of it on the page.

I may have held the PhD in Developmental Psychology, but Grandma Mary was the one whose experiment produced the findings worth sharing.

Early childhood experts often give this advice to new parents: talk, talk, and talk some more. This is true for families with deaf and hard of hearing infants as well; experts advise exposure to sign language right from birth or the time a hearing impairment is diagnosed. But the parent-child language connection has its beginnings before your child is even laid in your arms.

A child's exposure to spoken language begins before they are born. The auditory system functions by the start of the third trimester of prenatal development so that the developing child can begin to perceive the rhythm, pitch, and volume of speech sounds while in the womb. Once an infant is born, they hear the marvelous contours of spoken language with greater detail as they listen to their parents and other caregivers speak to them. They'll even recognize your voice from when they were in the womb, to know that you are a safe, trustworthy, and important figure in their life.

This grand conversation that begins between you and your child is a crucial one for many reasons. Language is how young children can convey their needs, connect socially, work through their emotions, enrich play, express creativity and imagination, and learn about the world around them.

And language serves a deep and vital function for your child's faith formation as well. Language is one of the main ways we teach our children what we believe. We tell them Bible stories, we say prayers, we explain why Jesus' teachings guide us today, and we express our own faith. Language is one of the most potent means we have to pass on faith to our children. Knowing about child development findings on language can help you focus on ways to deliberately introduce and reinforce faith concepts to your child during your everyday interactions together.

Wired for Language

Neurologically, babies are wired for language acquisition to occur right from the start, beginning with their well-formed auditory systems. Studies have shown that newborns are able to hear the difference between their mother's speech and another female speaking to them. A classic study from 1980 showed that newborn infants sucked harder on a pacifier when a recording of their mother's voice was played, compared to another woman's voice.

As their visual systems improve, babies begin to show preferences for what they look at while they listen, combining the senses of sight and sound to learn more about their world. Specifically, babies prefer round shapes to angular ones, color contrasts to sameness, and moving forms to stationary ones. All these preferences point your child to your face as you talk to them—they want to look at the round shape of your face, eyes, and mouth; at the contrasts seen in your hairline, your eyebrows, and your open mouth; and at the movements of your mouth, eyes, cheeks, and other parts of your face as you speak. Babies are captivated by our animated facial expressions as we talk. And likewise, we are drawn to their engaged, captivating baby faces, making us willing to slow down from the busy world around us and connect with someone who can't say a word but is still able to carry on a conversation.

Speak Parentese to Me

When we open our mouths to speak to our infants, something amazing happens. Without any training or special classes, we speak a new language—parentese! This language is mostly the same as the languages we speak to others around us, but with some definite changes. Our speech slows down, and we use simpler vocabulary and sentence structures. The pitch of our voice rises higher and is more animated. It is often paired with gestures, movement, and gentle touch. (Parentese is different than what's been termed "baby talk," in which non-sense words are spoken and repeated. While baby talk isn't harmful, it doesn't support language development the way parentese does.)

Studies have shown that infants prefer this form of language—the simple, animated speech of parentese—compared to typical adult speaking. One reason may be that parentese helps babies start to learn the breaks between words and the back-and-forth nature of communication.

The term *motherese* was used for many years but has been replaced by *parentese* to acknowledge the fluency that fathers and other primary caregivers also have when talking to infants. Not only do dads use parentese, but this form of speaking has been shown by siblings and other older children as well. Something deep within us knows that for babies to be in on the conversation, we need to make some adjustments to how we speak.

Critical Windows for Language

Early childhood experts use the term *critical window* to describe specific time periods when a child's brain is neurologically sensitive to the effects of specific experiences. Similar experiences at other times in a child's development will not shape neural pathways as significantly as during these critical windows. (Humans aren't the only ones among God's creatures to have critical-development windows; for instance, songbirds have a critical window in the first few months after hatching when they need to learn the love songs that will be crucial for mating later in life, and even shrimp have to be exposed to salt water as babies in order to survive into adulthood!)

Experts who study the critical windows for human language development have identified the first few years as crucial for exposure to language so that a child can become a fluent listener and speaker. Research indicates that for phonetic learning—the ability to hear individual language sounds and distinguish between them—the critical period occurs before an infant's first birthday. Most languages have about forty of these individual speech sounds, called *phonemes*. The word *ball*, for example, has three phonemes: the beginning *b* sound, the middle *a* sound, and the ending *l* sound.

One series of studies assessed this critical window by presenting phonemes in a foreign language to infants who had never heard the language before. Amazingly, babies could hear differences between these foreign language sounds that older children and adults could not discern. This means that your baby has the ability to become more fluent in foreign languages than you could ever be, and it helps explain the growing popularity of language immersion schools, preschools, and even day-care centers. As early language expert Dr. Patricia

Kuhl puts it, babies are "citizens of the world," able to learn any language spoken to them during the first year. However, their ability to hear this difference begins to diminish at about eight months, and by their first birthday, babies can no longer hear the differences in foreign languages. This research shows that one critical window for language learning—hearing differences between all phonemes—has closed by twelve months.

For learning syntax, or the structure of words and phrases, the critical period is between eighteen and thirty-six months. In lab settings, toddlers can detect sentences with inaccurate structures, or syntax. Some studies using brain mapping techniques have even shown that different neural pathways are activated when toddlers listen to incorrect syntax compared to sentences that follow the rules. These findings show that while toddlers still speak in short phrases and sentences, they can understand more complex sentences and detect when they don't "follow the rules," even though they've only barely begun learning the rules. Another critical window in language learning occurs later in childhood. If children learn a second (or third, or fourth) language before puberty, they typically speak it without an accent.

All this talk about critical windows might have you worrying that you've missed out on helping your child develop in a certain way, particularly if one of the windows has passed, if you have a child with a disability or developmental delay, or if you've adopted a child later in their life. But don't fear: it's never too late to help your child develop the skills they'll need later on. As we saw in chapter 2 on brain development, your child's brain is tremendously plastic throughout the first years of life. No matter who your child is or how they have developed in the *past*, you are the best parent for your child right *now*, and you can help them grow by doing what you're already doing— providing them with a safe, loving environment, including challenges and new experiences to help them grow! God is with you as you care for your child, and through your loving care, God is also present in your child's life.

Before a First Word, Millions Are Heard

Have you ever wondered how many words your little one has actually heard over the course of the days, weeks, months, and years that you and others have been speaking to them? One influential 1995 study sought to quantify the number of words children hear in the first years of life. For two and a half years, these researchers recorded spoken language for an hour each month in forty-two families with infants and toddlers. After recording spoken language in these households, the researchers used a statistical process called extrapolation. This allowed them to estimate how many words a child in each household would hear annually based on the home observations.

Their findings were astonishing, in part because of the discrepancies they found: while some of the children heard about 15 million words by their fourth birthday, those raised in more language-rich households heard about 45 million words by age four. Their study has led to a campaign called the 30 Million Word Gap that funds education and training programs to narrow this discrepancy. While we hope every child grows up in a language-rich environment, the sheer number of words heard is worth noting. Each of a child's first four years is marked by hearing millions of words, which breaks down to thousands of words daily.

For parents, the takeaway is very clear: talk to your child as much as you possibly can. Many of the millions of words your child hears in the first years of life will come from you! You can also encourage other caregivers—such as grandparents, aunts and uncles, and day-care providers—to talk to your child as well.

Receptive + Expressive

Researchers studying early language development have discovered many other remarkable features of language acquisition in infants, toddlers, and preschoolers. The work typically differentiates between *expressive language*, the words that a child speaks, and *receptive vocabulary*, the words that a child can understand. In general, researchers have found that children can understand a lot more than they can say, at every stage of development.

For instance, researchers studying memory in early childhood have learned that infants show that they comprehend common words before their first birthday, starting as early as six months. Babies understand what we say way earlier than we think!

In studies in which new words are introduced to toddlers, they're significantly more likely to remember them when they have cues and support about what these new words mean. Our toddlers need us to explain and describe new words; when we do, they remember.

Researchers have found that a child's ability to remember words steadily improves during the first two years of life. As children hear more words, understand more words, and speak more words, they remember more words.

Your child's first words signal a breakthrough because of how powerful a tool language is, allowing them to better communicate with others and express themselves. Their vocabulary may begin as just a few words for objects and people, but typically by eighteen months, language acquisition begins to grow exponentially in what has been termed the *language explosion*. A child's vocabulary of a few words grows to a few dozen, then a few hundred. By age three, their expressive vocabulary is typically between two thousand and four thousand words. For children experiencing speech and language delays, these numbers can be significantly lower, but their receptive vocabulary may not be as affected. If your child's language doesn't seem to be developing as quickly as in other children the same age, it's likely that they understand much more than they are saying, and are simply moving at their own speed. That said, never hesitate to consult with your child's pediatrician if you are worried. If there is a problem, many resources and language experts are available to help, and early intervention is often beneficial.

No matter the circumstance, bathing young children in language right from the start can form a foundation for language that supports healthy development in many areas, like socialization, emotional regulation, and academic success.

Introducing the Language of Faith

Research has shown conclusively that the first years of life are a remarkable and essential window of time for language learning. And because language learning is the foundation for so many other kinds of learning, these years are also a critical window for teaching faith. Knowing that your young child will be exposed to millions of words before their fourth birthday can be a thrilling discovery for you as a parent—but maybe a daunting one too. (Since when did parenting include having to think about millions of anything?)

Rest easy. The methods for building your young child's language skills—at the neurological level, no less!—don't involve special classes, vocabulary lists, or lengthy how-to tutorials. They do require you and other caregivers to engage in an ongoing dialogue with your little one using the conversation starters around you—the sun shining through the windows, the elephants printed on their pajamas, the milk in the bottle or cup, the silly antics of an older sibling or friend. The phrase *bathed in language* helps describe a child's total immersion in these kinds of language experiences. Ideally, we want every young child to be bathed in language—spoken to when they wake up, during diaper changes, while they eat, play, and get buckled into the car seat.

In your role as Chief Language Bather, you have many opportunities to introduce and reinforce a language of faith during your child's first years.

Language about God

In the previous chapter, we talked about how you model God's love for your child every time you provide physical care that is gentle, nurturing, and responsive to their needs. But you can also speak words about God so your child begins forming an understanding of who God is. Your language about God and faith will be shaped by your own personal beliefs and theology. Consider the following ways of using simple language about God with your young child.

God as Creator: God is the one who created the world. God's creation is good. Your child can learn that God is the one who made us all and that we are part of God's good creation.

God as Provider: God provides us with what we need. God gives us people who help provide for us. Your child can learn that their needs are met because God provides food, clothing, shelter, comfort, and loving caregivers who take care of them.

God as Protector: God protects us from harm. God is with us when we feel scared or lonely. Your child can learn that God is present even when we cannot see God, and we can practice prayers and songs that ask God for help and protection.

God as Forgiver: God forgives us when we sin. God will always give us another chance. Your child can learn that asking for forgiveness is an essential part of our relationships with others. Forgiveness wipes the slate clean so our relationships can be restored.

God as Parent: God cares for us as a heavenly parent. God's love for us is deep. Your child can learn about God's love for them through your parenting. We are created to be in relationship with God, who has an everlasting love for us.

Language about Jesus

Through storytelling, prayers, and conversations, your words introduce your child to Jesus. In some ways, this language may be easier because you can talk about an actual person who did things that we do today—walking, talking, eating with friends, and helping other people. These five descriptions of Jesus provide some language about how parents can talk with their little ones about him.

Jesus the Baby: Jesus was born to a human mother. He was just as vulnerable as babies are today. Your child can form connections to the story of Mary's pregnancy, Jesus' birth, and the wise men's visit to young Jesus.

Jesus the Healer: Jesus saw people who were sick or hurt and helped them feel better. His miraculous healing changed the lives of the people he touched and the people around them. Your child can learn that we ask Jesus for healing and wholeness.

Jesus the Teacher: Jesus taught people, from his inner circle to the great crowds that followed him. Jesus' teachings helped people know more about how they could love God and each other. Your child can learn that Jesus is someone who taught many years ago and who still teaches us now. Everyone can learn from Jesus, from little ones to grown-ups.

Jesus the Son of God: Jesus lived a human life, but Jesus was also God's Son. Jesus was sent by God to teach us how much we are loved by God. Children can learn that Jesus did many things to show us he was God's Son.

Jesus Our Savior: Jesus died on a cross and rose from the dead. The death and resurrection of Jesus form the cornerstone of our Christian faith. Your child can learn that these two events are important parts of the story we tell about Jesus.

It is not always easy to share our adult beliefs and theological views with young children. Sometimes we don't think we have the right words. Sometimes we'd rather change the diaper quickly without a conversation. Saying what we believe in language that is simple but not simple-minded can be a challenge.

Fortunately, we have many opportunities during our everyday activities with our children to use language about our beliefs so that they never remember a time when their caregiver didn't talk about their faith.

Little Steps

- What are the most important words of faith you want your child to hear every day? Write them down, post them in your home, and repeat them daily. (If you want to get fancy, find a DIY tutorial for how to make them into artwork for your walls.) Surrounding yourself with reminders can help prompt you to speak these words to your child each day. Posted words will also support your child's early literacy development as they see text that matches the words they hear. Here are some ideas for these daily words of faith:

  > You are a child of God.
  > I love you and God does too.
  > God's love for you is sooo big!
  > You are my precious child.
  > God loves every part of you.
  > Mama/Dada loves you sooo much.

- Weave faith language into your daily observations as your child's roving reporter, constantly describing the things going on all around you. Whether you're looking at the sunshine, talking about the weather, pointing out parts of your child's body, or sharing your feelings, you can use language that roots all of our lives into what God has created.

- Learn a few short, simple prayers and repeat them before meals, bedtime, or other daily activities. Babies hear the rhythm and sound, toddlers will be able to repeat many of the words, and preschoolers may begin to initiate these prayers themselves, even when you're not around.

- Write down your child's words when they begin to use faith language. As amazing as their proclamations will be in the moment, you will not remember them because other amazing discoveries will crowd them out of your memory! Use a journal, write them on sticky notes, keep them in the Notes app on your mobile device, or use any other way of recording events, but do write them down. You can weave these events into the stories you tell your child about when they were little, because big kids love hearing stories of their younger days. If your child becomes a parent someday, they will treasure having stories from their own childhood to inspire them and to pass on to their children.

- Read out loud from your own Bible to your child, or buy an age-appropriate children's Bible to read to them. Whether you already have a regular practice of reading from Scripture or are just starting to explore how to do this, your child will benefit from hearing your voice speak God's Word.

- Ask other faithful people in your family's life to share words of faith with your child, whether as a story, a blessing, or a prayer. Children benefit from hearing different ways that trusted loved ones express their faith. Whether in person, via a video call, or even through a voice recording, your child can learn that their loved ones speak words of faith.

5

Literacy:
Introducing Stories of Faith

One of the most thrilling and demanding projects I've worked on during my publishing career happened when my team developed the Spark Bible, a full-text Bible for upper elementary kids. This Bible was designed for them to write comments in the margins, add stickers, and mark up passages to show how they encountered stories in their Bibles, whether they were in Sunday school, at home, or on the go.

The release of this Bible was big news for our company. Our President and CEO, Beth Lewis, sought to emphasize the importance of the new Bible by speaking and writing about a Bible from her past—specifically, the Bible she'd received at age six or seven from her grandparents. Bound with a black leather cover and a gold metal zipper with a dangling gold cross, its onion skin pages could easily rip if turned too quickly or roughly.

The gift of a Bible from Beth's grandparents spoke of the importance of passing down faith. It also sent a powerful unspoken message to its young recipient: approach the Bible with caution, don't get too comfortable with its pages, and keep it serious. She treasured this gift from her family, keeping it for decades amid many moves and transitions. But Beth celebrated the release of the Spark Bible in a way that marked that something new and different was happening with children and God's Word. Our company could now offer a Bible that would invite children to actually read its pages, then add notes, drawings, stickers, and other ways of questioning, commenting on, and engaging with the text. By holding her childhood Bible side by side with the bright orange Spark Bible, she was emphasizing that the very experience of reading from God's Word could— and should—be a markedly different one for kids today.

Before exploring what *supporting early literacy* means and what it is, it's important to clarify what it *isn't*. Supporting early literacy in your child doesn't mean planning activities so that they will begin to read earlier than their peers. (This means you can put down those flashcards, uninstall those twelve learn-to-read apps from your mobile device, and breathe a sigh of relief.) Instead, early literacy is an expansive approach to designing experiences for your child so they are exposed to print in many ways across many settings. Learning to read is the culmination of a wide variety of experiences—hearing stories, handling books, writing letters, looking at illustrations, and, very importantly, seeing how loved ones value reading and share reading time with them.

Learning to read is also critical to faith development, since so many of the core beliefs and practices of the Christian faith are centered on or passed down through the written word. Christian thinkers and seekers throughout history have sought to understand God better by turning to writing and reading, and one of the central objects of our faith is a book—an inspired, holy book called the Bible. Let's dig into the research on childhood literacy and how this research can help you prepare your child for a lifetime of joyful engagement with God's Word.

The Importance of Reading

Reading is a crucial life skill for children, youth, and adults. Few abilities are as strongly linked to so many later outcomes, beginning in childhood. By the time children enter kindergarten, they'll display a few strong predictors of reading success, including knowledge of the alphabet, ability to hear and distinguish letter sounds, and the capacity to write letters, including their names.

By fourth grade, a child who is a fluent reader is ready for what reading researchers call the transition from *learning to read* to *reading to learn*. Literacy at this age is critical so that children can read about subject matter such as social studies and science in addition to what they learn during additional reading instruction. Reading proficiency during this window is linked to academic success in high school. Low literacy in adults is linked to significantly higher rates of poverty, underemployment, the need for government assistance, incarceration, and several other bleak outcomes. A newer area of study called *health literacy* points to the importance of reading skills to ensure that adults can navigate the health care system, understand health information, and manage medical conditions.

Reviewing statistics about the devastating effects of low literacy and illiteracy can be overwhelming. Here's the good news: parents and other caregivers of young children can plan early literacy activities with their children that are enjoyable, easy, and effective. These activities can help your child develop key skills that are linked to success in learning to read—even if they're only an infant or a toddler!

The National Early Literacy Panel has identified several skills during the first five years of life that predict later literacy development:

Alphabet knowledge is defined as knowing the names and sounds of letters. Your child may sing along to the ABC song, calling out a few of the letters at first, then singing more and more. Your child may recognize their first initial, whether it's hanging as a decoration on their bedroom wall, in a book, or part of a corporate logo. Alphabet knowledge is usually shown in visual ways.

Phonological awareness is a skill that builds on the ability to recognize individual sounds, called phonemes. It requires that children hear letter sounds, distinguish between different sounds, and show they can segment words into sounds. When your child identifies the sounds that begin or end a word, or when they can predict what word ends a line in a nursery rhyme, they are showing this early literacy skill. Phonological awareness is auditory. Children show their levels of awareness through what they can hear.

Letter writing is shown when children write letters of the alphabet using crayons, markers, pencils, or other writing utensils. This skill is on display when your child begins to write the letters of their name. Young children may write random letters or use what's called *invented spelling* to write words. This early literacy skill requires fine motor skills developed enough to write letters.

Print knowledge can be seen when children show they know how to use and interact with print. When children look through board books on their own, hold the book right side up, and flip through pages from front to back, they are demonstrating print knowledge. They also show this knowledge when they notice print in other places, such as a road sign, restaurant menu, or newspaper.

Oral language is measured by a child's comprehension of the words spoken to them (receptive language) and the words they use (expressive language). Children's oral language skills can be assessed by their understanding of words spoken to them and the words in their vocabularies.

Experts who study literacy have developed precise definitions and measurement tools for each of these skills. They've conducted long-term studies to identify the most significant contributors to literacy development. They report their findings in journal articles, books, and conferences. Fortunately, they've also made sure the results from these studies don't stay in the universities, nonprofit organizations, and other institutions that led and funded their studies. Literacy experts want their research to help parents, educators, and others create language-rich environments in homes, day cares, preschools, libraries, and other early childhood settings. Your local accredited early childhood center likely has access to research-based tools that help assess and build early literacy in child care and preschool centers and in homes. If you ask an early childhood educator, "What can I do to help my little one build literacy?" expect an enthused (and maybe lengthy) answer.

Faithful Parenting and Faith Literacy

Your child's literacy development begins long before they name letters, read words, or even speak first words. You can design a language-rich environment for your child that supports their literacy in age-appropriate—and fun!—ways. What you may not have considered is how faith formation can result from your efforts to build your child's language and literacy skills. Talking with your child, reading to them, and surrounding them with faith-based books and other print materials can serve a dual purpose: building print literacy and introducing them to ideas about your Christian faith.

A book for all ages and stages: A foundational part of faith is hearing and remembering stories from the Bible. Reading to your child from a children's Bible builds print literacy and introduces your child to a book that can accompany them throughout their entire life. The Bible helps us answer the big questions of life, beginning with major developmental issues that infants and toddlers face long before they can put these needs into words: Who am I? Who takes care of me? Will my needs be met? Who can I trust? Who loves me? Who do I love?

Bible stories address many of these questions that infants and toddlers have. We read the creation story and God's words, "It is good." We tell our children about the rainbow's colors, reminding us of God's promise to Noah. We delight in how Moses' clever older sister ensures he'll be cared for by family. New Testament stories offer words of comfort and assurance about Jesus' love and care for us, along with stories about broken relationships that are made right through repentance and forgiveness. The themes, ideas, and narratives we read in Scripture offer us guidance, wisdom, and insight as we live out our lives of faith. But there is something for our young children in these stories as well when we trust that the stories themselves are meeting needs children have for hearing stories that are powerful, well crafted, and relevant to their lives.

The variety in the Bible: Reading experts advise parents to read all kinds of books to their young children so that they will learn how books tell different kinds of stories. When we read stories from children's Bibles, we also give our little ones the opportunity to hear different formats, emotional tones, and storylines. They get to know long names and short names, familiar names and unusual names. They hear dialogue between God and the people called by God. They find out about exciting stories of danger and adventure, along with quiet, reassuring stories about loving care and security. They hear the poetry of Psalm 23 and the rhythmic language of the Beatitudes. And they witness the emotional highs and lows of God's people, who experience life and death and resurrection.

The "bookness" of the Bible: Early literacy best practices involve exposing infants and toddlers to physical books, text on the page, reading aloud, the basics of storytelling, and the use of illustrations to support storytelling. These activities help young children develop the concept of *bookness*, which refers to the qualities that make up a book, and the experience of interacting with a book. Your child will expand on this concept as they grow older, hear more stories, and recognize more about the world of the printed word. When the Bible is one of the books that helps develop this concept of bookness, children never remember a time when they weren't engaged physically with God's Word.

Print literacy + biblical literacy: Hearing Bible stories and looking at illustrated children's Bibles both help your little one develop print literacy. These activities also start building the foundation of biblical literacy, defined as basic familiarity with the different parts of the Bible, such as the specific books and the two testaments, and the ability to find specific verses within chapters. Your child has many years to go before they read their first words from a Bible or know the names of the Bible's books and the timeline of biblical events. But reading Bible stories to them now sets you on a path for both print and biblical literacy to occur.

Churches that start early: If you are part of a church community already, what do you know about your church's approach to supporting early faith formation through literacy activities? (If you are not yet part of a faith community, chapter 9 of this book is waiting for you!) Churches can be warm, welcoming places of faith to young children and their parents through the ways these communities greet families, care for children, and equip their spaces with age-appropriate toys and furnishings. A well-stocked children's library is one sign of a church that is tuned in to the needs of little ones. Churches can also support the development of biblical literacy in babies and toddlers by planning times for their caregivers and teachers to read Bible stories to them. Imagine if every child could never remember a time when they didn't see and hear God's Word read aloud to them by their teachers at church! The church is a vital context that supports early literacy by introducing the Bible as a book that is important to read to each child because it is important to our faith.

Little Steps

- **Read words (and the Word):** Literacy experts and early childhood educators share this recommendation: read to your child every day. With an abundance of story Bibles available, you can include Bible stories as part of your daily reading time with your little one. Bibles for young children have brief stories you can read quickly in the face of short attention spans (your child's, or maybe your own), along with inviting illustrations.

- **Building print literacy can build biblical literacy:** For Christians, the sooner a child develops biblical literacy, the sooner that child will be able to go beyond individual stories in Scripture to see broader themes and ideas expressed throughout the Old and New Testaments. Reading to your child from children's Bibles and your own Bible shows that your family reads about God from the pages of a book we call the Bible, and that there are versions for little kids, grown-ups, and everyone in between.

- **Rhyme time:** Rhymes are structures to help children develop phonological awareness. Rhyming stories, poems, and songs can be tools for sharing playful words and emotions. Bible verses, stories, and prayers written as poems or as songs with rhyming lyrics can reinforce this important skill with your child.

- **Build your library:** When choosing books for your child's library, make sure to include age-appropriate Bible storybooks. But don't stop there! Add books about faithful Christians and stories about values and beliefs to your collection so your child has an array of books to choose from that reinforce practices such as telling the truth, asking for forgiveness, and sharing with others.

- **Talk the talk:** Spoken language is a strong predictor of literacy. Begin by speaking faith-based words to your child (see the previous chapter!), then read books about these concepts. When you talk about telling the truth, asking for forgiveness, or sharing with others, follow up by reading stories about these important values. They can be Bible stories, folk tales, fables, or other narratives that reinforce these ideas.

- **Writing down faith:** Writing skills also contribute to literacy development. As your child learns to write letters in their own name, pair your child's name with a blessing or affirmation. Teach the letters *G-O-D*, or make *J*s together as you make the sound and pair it with *Jesus*.

- **Seeing what words matter:** What are important words of faith to your family? This may take some time for families to figure out, especially if you've never considered this question before. Write them down, make art of them, post them on a Word Wall, or figure out other ways that get words that matter into clear view of your child. Some families even write what they call a family creed, a statement of belief about your family that lets everyone know who you are and what you value. Such a statement is a way to echo the beliefs we profess when we say the words of the Apostles' Creed.

6

Music:
Your Playlist
Makes a Difference

Years ago my husband and I played music for a friend's wedding. During the rehearsal at the church, I noticed a toddler hovering close by his mother who was the pianist for the service. The same little boy captured my attention as the congregation sang on the day of the wedding. (Since I played accordion at the front of the church for all the hymns, I faced everyone as they sang.) This little boy stood in his dad's lap and opened and closed his mouth while everyone sang together. The movements of his mouth weren't random, and his attention didn't fade after the first verse of the hymn. It really looked as if he was following along with the music and lyrics, just like everyone else, even though he was only a few years old.

After the service, I asked his mother about what I had observed. She told me how her little boy had been immersed in music since he was in the womb. Because she was the choir director, he was surrounded by music for hours each week during the nine months of her pregnancy. After he was born, she brought him to midweek choir rehearsals and Sunday morning worship services. Once he could start walking, she discovered that he would stay seated (usually) so that he could hear the music they were making in the choir. "He sings along, even though he doesn't know the words," she reported. "But he knows he's part of the group. He'll learn the words soon enough."

Most parents have witnessed their baby connecting to music in an immediate, physical way. Have you seen the calming power of music soothe your baby for naptime or nighttime? Or maybe you've enjoyed the way a lively tune leads your child to moving their body in an animated, rhythmic way. Once children can stand and walk, their response to music becomes even more delightful, as they bob up and down, sway side to side, raise and lower their arms, and make sounds with their voices that approximate singing. Young children already know music has power to move us, to teach us, and to connect us—and they want to get in on the action! What's more, because music can be an irresistible way to invite children into movement, singing, and community, it gives us a captivating way to introduce faith to our little ones.

The Music of Early Childhood

For several decades researchers have studied music perception in infants. The results have repeatedly shown how humans are capable of actively listening to and making meaning from music. Using an array of measurement techniques, researchers learned the many ways babies are capable of perceiving music's features—tone, melody, tempo, emotion, volume, and more.

But far from being passive recipients of music, young children are equipped with abilities and preferences for music. They show their interest through their gaze, movement, facial expressions, and even changes in heart rate and other biological signs. More recently, brain imaging technologies have shown that specific regions of the brain light up when children listen to music. And some studies demonstrate that babies can detect differences adults can't pick up on between the music of various cultures.

Following are just some of the ways in which little ones are already astute music listeners and participants in their first years of life.

Harmony over cacophony: Babies prefer music that sounds harmonious rather than dissonant. In one study of four-month-olds, children listened to a sequence of consonant and dissonant melodies. (If you know about musical intervals, imagine a tune made up of major thirds compared to minor seconds. It's like the difference between Mozart and the theme

from *Jaws*.) Infants looked longer at the source of the consonant songs compared to less harmonious ones. When the dissonant songs played, they turned away from the sound more
often, and their actions were more fretful.

Moving to the music: Researchers have also looked at the
movements of babies and toddlers when they hear music. Not
surprisingly, young children move more rhythmically to music
than to speech. They also adjust the speed of their movements
based on the tempo of the music.

Waltzes help talking: When infants were exposed to songs
chosen with distinct time signatures (for instance, the 3/4
count of the waltz), measures of their brain functioning showed
that they were able to detect and predict the tempos of other
waltzes *and* of speech more effectively as well. These results
suggest that when young children listen to music, they may
be forming neural pathways that help them recognize speech
patterns more effectively, applying learning beyond the realm
of music.

Music interventions at work: Participation in music therapy
programs during a child's first few years of life has been shown
to have positive results later on, including an increased ability
to engage in play with parents and peers.

This sampling of research into children and music demonstrates that young children are capable of perceiving music in
more complex, nuanced ways than we may have ever imagined.
Just as children are wired to begin making meaning from the

words we speak to them, they are also neurologically designed to hear and begin to understand the language of music. In the presence of music, a young child is not a passive listener. Instead, infants and toddlers have brains that are ready to hear music, make meaning of its structure, and form new neural pathways in response to a range of musical experiences.

Parenting with Song

What do these findings mean for you and other caregivers of your child? How can you apply them to your busy life of parenting? And how can music support your child's faith formation? Just think of how central music is to the Christian faith. During worship, we sing songs to internalize the messages and meanings of Christianity. When children gather for Sunday school, the first ideas they get about faith often come from simple but powerful songs such as "Jesus Loves Me" and "This Little Light of Mine." Long before the scientific studies telling us about the power of music, surrounding children with "psalms, hymns, and spiritual songs" (Colossians 3:16) has been one of the most tried-and-true ways of passing faith on to kids.

Connecting Music to Scripture

To see the importance of music to faith, we need only look to the Bible, where some of the most powerful stories of God's actions in human lives are accompanied by songs. From the beginning, God's people have testified to the divine presence in their lives with music. Here are just a few examples:

- One of the earliest mentions of music in the Bible is in Exodus 15, when Moses, and then his sister Miriam, lead the Israelites in song after they crossed the Red Sea into freedom. In verse 20, we read that Miriam and other women accompanied her song by playing tambourines and dancing.

- Before he was king, David played the lyre and sang for King Saul. Many Bible scholars consider David the composer of a great number of the psalms, along with songs found in 1 and 2 Samuel.

- The book of Psalms is full of song-poems that express the rich range of human emotions. Many have been set to modern music so we can imagine what the psalmists may have been thinking, feeling, and even singing as they penned the words thousands of years ago.

- Song of Solomon has the word *song* in its very name. It is a book full of love songs. While you may be more likely to sing these to your spouse than to your baby or toddler, it reinforces to us that throughout the millennia, God's people have used music to express the deepest parts of themselves in relationship with others.

- Jesus' mother, Mary, sang a joyful song of praise while visiting her cousin Elizabeth during her pregnancy. The words of this song, known as the *Magnificat*, convey her praise and reverence to God.

- Jesus and his disciples met for the Last Supper, then sang a Passover hymn together before leaving for the garden of

Gethsemane. In the midst of this fraught time, a familiar song united them.

- When they were imprisoned, Paul and Silas stayed up late to pray and sing songs, until an earthquake disrupted their midnight praise session.

Throughout the ages, composers, lyricists, and musicians have expressed their Christian faith through music, whether the lyrics were based on scripture or were other words of faith. Music has the power to express the depths of who we are and what we believe. We are designed to listen to, enjoy, and make meaning of music. This deep connection to music begins when we are infants. As parents, we choose the music our children listen to for the first several years of their lives, until they learn to make their own playlists.

The Best Voice Is Your Voice

If someone asked you about the music in your child's life, how would you respond? Maybe you'd talk about music formats, like your favorite streaming service, an affinity for LPs, or your old-school CD collection. You might recommend lullabies sung by pop stars or simple songs recorded by musicians specializing in children's music. But nothing on your playlist compares to the songs you sing to your child. God gave you the gift of a voice as a just-right way to connect with your child through song. Following are just a few of the many ways singing to your child can benefit both of you:

Language enrichment: Singing introduces language concepts in engaging ways. Before babies speak their first words, they tune in to the contours of speech, including pacing, pauses, and inflections. The songs you sing give them familiar and engaging ways to experience language. Songs can also introduce toddlers to new vocabulary in exciting—and sometimes humorous!—ways by giving context clues for new words and repeating them.

Sing to me, Mama! (And Daddy, and Older Sibling . . .): You don't need to be a singing pro to serenade your child. However you sing, research shows that caregivers know how to appeal to their audience of one. In one study with mothers and infants, researchers asked mothers to sing a lullaby ("Rock-a-bye Baby") or a playsong ("Itsy Bitsy Spider") to their baby and then sing the same tune without the child around. They collected video and audio recordings of the mothers during these serenades and found that mothers adjusted their singing in noticeable ways when they had their baby as an audience. They paused longer between phrases, sang the song at a higher pitch, and showed more emotion in their singing. Those who coded the results reported that when mothers sang to their babies, they had a more loving, smiling tone of voice than when their baby was not present. Similar results have been found when asking fathers and older children to sing songs with baby present or absent.

You provide customized playlists: You know your child's preferences and moods the best. Whether they need reassurance, excitement, or a source of distraction, you can pick the songs that best fit the moment—and change to a new track or artist a few moments later if needed. The more songs you learn, the better able you will be to choose one that matches their moods, whether silly or sad, excited or exhausted.

Sing in community: Recent studies examining the neuroscience of singing have found that singing together in adult groups leads to the release of more "feel good" neurotransmitters such as oxytocin and serotonin. Many parent-child classes include singing, dancing, and movement to music. When you give your child a chance to be part of a broader group that sings together, you're supporting healthy brain development for them—and for you!

Little Steps

- **Put church on Repeat.** The music your family hears while joining others to worship in a church community can be a powerful way for your young child to connect to faith language. If you can access recordings of the music sung and played at your church, put them on your at-home or on-the-go playlist.

- **Sing faith songs during transitions.** Music helps signal to your child that a change in activity or location is coming. When you pair transitions with a song that affirms that God is with us during times of change, you teach your child about God's presence during the in-between times too.

- **Teach new faith words through music.** Singing is one way for children to learn new words. Introduce words such as *grace*, *forgiveness*, and *blessing* through song so that children hear them sung before having longer conversations about what they mean.

- **Sing your prayers.** Whether you compose a prayer song for your family, make up a new melody each time, or use a familiar tune, singing your prayers helps your child pair melody with words of praise, thanks, and asking God for help.

- **Get close.** Singing to your child as an infant and with your child as a toddler and preschooler is a meaningful way to bond and develop closeness while you introduce words and melodies.

- **Build your repertoire.** If you don't know many simple faith-based songs, search online, ask a teacher at a Christian preschool or day care, or ask your friends what they sing. By sampling a range of songs, you can choose the ones that best fit your voice and comfort level.

7

Whole-Body Parenting: Caring with Touch and Movement

Rachel and Mike's first child, Izzy, was born at twenty-five weeks and one day gestation after a fraught pregnancy. They felt terrified, with so many questions: Would their daughter live? Would she be healthy? What if they weren't able to bond? To help soothe their fears, their doctors and nurses recommended kangaroo care: physical contact with an infant wearing only a diaper and being held on a parent's bare chest while both are covered with a blanket.

During these times of skin-to-skin contact as she held Izzy on her chest, Rachel witnessed the power of touch. "When Izzy was resting on my chest, her heart rate would stabilize, her oxygen needs would decrease, and her apneas would stop. She would be hearing what she should have been hearing in the womb—my heartbeat—instead of the whoosh and dings of the ventilator."

Kangaroo care affected Rachel too, as she wrote to a friend during this time. "Kangaroo care is like nothing I have ever experienced. Almost immediately I feel all weird and sleepy. She's so warm and wiggly and her CPAP makes a calm whooshing noise. The next thing you know, we're both kind of nodding off. We hold her for an hour or two at a time, but not much longer than that right now. We don't want to stress her out with too much stimulation."

Looking back at those first months, Rachel credits kangaroo care with allowing her to bond with her daughter. "That deep, quiet time between the two of us made up for the time she should have been in utero, kicking me in the bladder. It gave me the space to open my heart to this tiny creature, no matter what the future held."

Izzy eventually grew strong, and Rachel now shares, "Izzy is still a very snuggly kid, and we talk about how she was so tiny she could fit inside my shirt. When I hold her on my lap now and feel how gangly she is, it's a wonderful reminder of how far we've come and makes me so thankful for that. And sometimes, it's really helpful to have that reminder— as any parent knows!"

This book has focused primarily on your child's development, highlighting ways that scientific findings can help you better understand how to integrate faith formation into daily life with your child. This chapter includes an additional focus—what is going on in your *own* brain and body as you apply yourself to a daunting new role: that of parent to a growing child.

Parenting demands a set of complex cognitive skills. We often describe our approach to parenting in terms of how we think and feel about our roles as parents. Our emotions, beliefs, and values shape our parenting philosophies. Reflections on our own childhoods give us insights about family history and parenting patterns. We rely on memory to recall information that helps us in our day-to-day parenting. All these cognitive processes add up to make parenting a challenging, complicated task.

Multiple parts of your brain are activated while you're performing everyday parenting tasks. Take the **prefrontal cortex**, for example. This brain region is responsible for a set of skills called *executive functioning* that consist of several interconnected capabilities. Here is a partial list of executive functioning skills, along with an example of how you might use each one in the course of a typical day as a parent to a young child:

- Decision making: *Should I let my child fall asleep on the ride home if it means ruining the chance for a nap?*

- Focused attention: *I need to read this book to my child, point to pictures, and ask questions, then do it again.*

- Planning and prioritizing: *If I warm up yesterday's noodles for lunch and serve applesauce instead of cutting up fruit, we can eat lunch five minutes sooner.*

- Emotional regulation: *I felt like bursting into tears when the store didn't have the right kind of formula, but I held it together.*

If we view parenting primarily as a complex set of cognitive and emotional activities, then our views on child-rearing stay primarily in the mental realm. Such a perspective views our body as simply the means for carrying out parenting tasks—the vessel that carries our brain around and does what it says. But we can go beyond this understanding if we embrace a deeper purpose for our *body* as we parent our children, being present for our kids with more than just our brain. I have started using the term *whole-body parenting* to describe this approach of including our whole physical selves during our daily lives as parents. I envision whole-body parenting as involving three specific areas—breath, touch, and movement—physical actions that are also strongly connected to faith.

As I summarize some research on each area, I invite you to think about parenting as a calling of the whole self, both mind and body. And here's the bonus—when you focus more deliberately on these physical expressions of parenting, you can also teach your child about them, amplifying the effects!

Basics about Breathing

Before reading this section, close your eyes and take three deep breaths. Try inhaling for several seconds, holding for several seconds, and then exhaling with a whooshing sound.

Done? You just performed a breathing exercise that can help slow your heartbeat and reduce stress.

Breathing is an action we do so regularly that we usually neglect this vital function. The act of breathing is usually automatic and involuntary. Most of the time we don't think about our breathing; a structure in the brainstem called the **medulla** regulates the ongoing act of breathing so we don't have to focus on it. If you know someone who has struggled with respiratory illness or disease or if you've experienced this yourself, you're aware of what happens when this bodily function is not as automatic and smooth.

Breathing is the only automatic function of our bodies that we can regulate voluntarily. (Try instantly changing things up with digestion, heart rate, or hormone release. Can't do it!) When we choose to focus on our breathing, the medulla yields control to the **cortex**, and we're now able to manage its depth and rate. You may have experiences with deliberately controlled breathing if you have played a woodwind or brass instrument, sung in a choir, or practiced yoga. "Use the breath!" or some variation of that command is shared by band directors, choir conductors, yoga teachers—and labor and delivery nurses.

In addition to these musical and meditative situations, controlled breathing can be helpful when we are feeling stressed or worried. Anxiety and tension activate the body's sympathetic nervous system, triggering responses such as increased heart rate, increases in blood pressure, and more rapid breathing. When we become aware of our breath and choose to control it so that it becomes voluntary, deep breathing can flip a switch in our nervous system. The parasympathetic nervous system, typically activated during "rest and digest" activities, counters the "fight or flight" responses triggered by our sympathetic nervous system. When we decide during stressful moments to take control of our breathing, we actually override those physical stress responses. You can't be in both states at once; when you start deep breathing, your parasympathetic nervous system tells your sympathetic response, "Knock it off—I'm in charge for now."

Research on breathing interventions has shown several positive effects, including lowered levels of cortisol, a stress hormone; reduction in stress, anxiety, and depressive symptoms; and increases in sustained and focused attention. And here's that bonus I mentioned earlier: while most studies have used adults as research participants, a growing number of researchers are examining how breathing interventions can lead to helpful outcomes in children as well. When you focus on breathing to support a whole-body approach to parenting, the positive effects can multiply. Deep breathing is a strategy you can teach your child too!

Touch

Recognizing the importance of physical touch is another way we can view parenting as a whole-body calling. Connecting physically with our children begins when we see and hear our little one for the first time. We long for this child to be placed in our arms so we can experience them with all our senses. We tune in to their appearance, the sounds they make, the way they smell! We also begin learning about them through our sense of touch—how it feels to touch their hair, their little fingers and toes, and their soft, squishy bellies. As adults, we have millions of touch receptors on our skin, with different ones feeling pressure, pain, temperature, and other sensations. Touching our young child activates our bodies and brains to begin forming connections that recognize the exact feeling of contact with him or her.

Our children are ready for this touch! Of our five senses, touch is the most developed one in newborns. They will need many experiences with gentle touch to form neural connections that recognize what human touch feels like versus other types of touch. Young babies also need regular physical touch to learn differences between pressure, temperature, and texture. For some children, some kinds of touch are not part of comforting interaction. Whether a child has a diagnosis of sensory processing disorder, autism, Asperger's syndrome, or another condition that affects development, their parents have a puzzle to figure out—what types of touch will help their child.

If you've had a little one who needed more time in the hospital after birth, like Izzy at the beginning of this chapter, you're already familiar with the power of touch. Skin-to-skin contact between premature infants and their caregivers has been studied for decades by Dr. Tiffany Field and her colleagues. Their work demonstrates that infant massage and kangaroo care help premature infants gain more weight, and result in their being released from the hospital days earlier than premature infants who do not have these experiences. When parents can be the ones providing this touch through massage and kangaroo care, they receive assurance that their touch has been therapeutic and helpful for their child's growth. That's using your body to parent for a great purpose!

(It turns out humans are not the only living creatures that suffer when they are isolated from touch. Labs studying touch-deprived baby rats find that they grow up to be more timid and high-strung, with higher levels of stress hormones. Touch even makes a difference with worms. When they are raised in isolation instead of in colonies where they can receive worm-to-worm contact, they were smaller in size and less reactive when their petri dishes were tapped. Poor little worms!)

As our babies grow and develop, we have countless opportunities for touch each day, but many of these are built into the care we provide through diapering, washing messy faces, or picking them up to put them down for a nap. In contrast to this task-based touching, *non-contingent touch* refers to the types of touch we have with our child that don't have a function. While the name might seem clinical or impersonal, it actually refers to the deeply personal types of touching that we do in families: snuggling, gentle hugs, and all the other ways that parents and children physically connect with each other *just because*. We want to make sure this type of touch is plentiful in our lives together because touching benefits both our children and us!

Get Moving

Whole-body parenting can also include movement with our young children. Babies begin their lives with many hours of stillness as they sleep in one position, but they are ready for movement. Starting at birth, a child's vestibular system is already well developed. This system consists of several structures in the ear that can detect gravity and movement. From the start of infancy, children already have a sophisticated sense of balance and motion perception. That may be why rocking, riding in the car, or sleeping in a moving cradle work so well—these are all activities that stimulate their vestibular system.

Your little one is dependent on you for movement experiences until they begin moving around on their own. As they become able to sit up, crawl, cruise along furniture, then walk, they are learning about the world around them. For them, moving is learning, and their bodies become a tool for learning. Sometimes you may feel like *your* body is on the move with the sole purpose of following your child and making sure they don't climb, crawl, or toddle into danger.

Parenting With and To the Whole Body

The concept of whole-body parenting can give you a helpful lens into faith formation as well. In Luke 10:27, Jesus calls us to love God and our neighbor with all our heart and soul and strength and mind. By naming our physical identities (our strength) in addition to our emotional and cognitive selves,

Jesus invites us to think about the ways our bodies express love. What better way to do this than through how we parent our children?

Use the Breath (God Did)

We know that breathing brings about some differences in our physiology almost immediately. We are created by God as living, breathing creatures who depend on the air that surrounds us. Scripture offers us many examples of the power of breath. Here are just a few passages:

- In Genesis 2:7 God breathes life into humans. The Hebrew word used in Scripture for "breath" is *ruach*, which can also mean "wind" or "spirit." In the New Testament, Acts 17:24-25 refers to God giving us life and breath.

- In the midst of his troubles, Job credits the breath of the Almighty for giving him life in Job 33:4.

- The prophet Ezekiel's vision describes God breathing life into dry bones in Ezekiel 37:5.

- After he has risen from the dead, Jesus breathes on his disciples and tells them to receive the Holy Spirit in John 20:22.

Breath gives life and breath has power. When we frame the act of breathing not just as an involuntary, automatic function of our physiology but as a life-giving gift from God that reminds us of our origins and of God's power, we can view breath in a new and powerful way while we parent.

Consider a few of the ways that parents notice, celebrate, and worry about breathing in their children during the first years.

- Baby's first breath: For parents who are in the delivery room when their child is born, the marvelous sound of their baby's first breath (and cry!) is a cause for celebration. When baby's breath is labored or troubled, it leads to worry and anxiety.

- Sleeping like a baby: Watching the rise and fall of a baby's chest during sleep is one of the delights that parents enjoy. Our infant's calm, even breathing assures us that all is well and that they are resting comfortably. (Of course, when their breathing becomes rapid, uneven, and even noisy during certain phases of sleep, things aren't quite as idyllic!)

- When breathing is hard: Hearing our little ones struggle with breathing because of a cold, cough, or other illness puts our parenting empathy on high alert. When our children can't breathe with ease, we notice and we want to help.

- Watch a toddler throw a tantrum and you'll see the "fight or flight" system kick in as their breath quickens, their heart races, and their face grimaces in tension. Some little ones even hold their breath in the midst of stressful situations.

A child's breathing gives cues about their inner state and their immediate needs. But as parents, we can benefit from focusing on our *own* breathing as well. Have you noticed your breath when you are stressed out, whether because you feel overwhelmed by parenting or are in the midst of disciplining your child? Your own "fight or flight" response may be acti-vated because the demands of parenting young children make us intensify our efforts and parenting strategies to match the situation. That's why when we learn to slow down and breathe together, everybody and every *body* benefits. Your child learns a way to calm down and self-regulate. You help your body do a reset so that you can parent from a calmer state of mind—and body. Breathing is essential for us to survive, but it can also help us thrive.

The Power of Touch

When we look to Scripture, we see remarkable accounts of people who are changed through touch:

- In Jeremiah 1:9, God used touch as a way to call the prophet.

- Isaiah 6:7 describes the call of Isaiah when a seraph touches a live coal to his lips. (Ouch!)

- In Daniel 10:10, 16-18, when Daniel is called, he is touched three times by a messenger of God: once to rouse him from a trance, once on the lips, and a third time to give him strength.

And we see the power of touch in the life of Jesus:

- Matthew 9:18-25 describes Jesus on the way to the house of a girl who has died. Her father has asked him to lay his hands on her. On their way, a woman suffering from bleed-ing touches Jesus' cloak, and he knows immediately that he's been touched. He sees the woman and makes her well. Then he goes to the girl's home and takes her by the hand—and she rises!

- Later in Matthew's Gospel (20:30-34), two men who were blind call out to Jesus and ask him to open their eyes. He touches them, and they regain sight—and follow him.

- While Jesus and his disciples are in the garden in Luke 22:51, someone in the crowd strikes a soldier with a sword and cuts off his right ear. Jesus touches the soldier's ear and heals him.

Other stories about Jesus show touch used for purposes besides healing: for comfort, affection, blessing, reassurance—the very reasons parents often touch their children:

- Matthew 17:6-7 describes Jesus' Transfiguration. When he is on the mountain with Peter, James and John, they are terrified. Jesus touches them and says, "Get up and do not be afraid."

- In Mark 10:13, parents bring their children to Jesus so that he will touch them. After telling his disciples, "Let the children come to me," he takes them in his arms and lays his hands on them and blesses them.

- John 20:24-27 tells us that Thomas will not believe until he touches Jesus' hands and side. Jesus invites him to do so.

The power of touch is amazing. As caregivers of little ones who are literally "in touch" with them each day, you have the power of showing them love, gentleness, and healing. We may think that our faithful parenting is all about what we tell our young children about Jesus, but a powerful way you show the love of Jesus is through your touch. And touch transforms us as parents as well. By sharing the gift of touch with our children, we connect with them more deeply and are empowered to be the kind of fully present, whole-body parents we are called to be.

Movement

Some church communities expect children to stay quiet and subdued during worship and other church activities, even though such expectations are not developmentally appropriate. (Read more about this in chapter 9.) Psalm 46:10 may call us to "be still and know that I am God," but in these early years, faithful parenting is not about reading the Bible quietly, memorizing scripture verses, and praying while everyone is calm. (For most families, those things happen maybe 3 percent of the time during a child's first three years of life. Or maybe 0.3 percent!) Using a whole-body parenting approach to our children's faith formation can incorporate movement in many ways. Children learn deeply by experiencing the world around them through movement. Forming their foundations of faith shouldn't be any different, but they may need to look to parents for ways that faith is expressed through actions, not just words.

Let's look first to Scripture and the many ways movement has been used to express faithfulness in our relationships with God.

Dancing to faith-based music has a long history:

- In Exodus 15:20, Miriam and the women dance and play tambourines after the Hebrew people cross the Red Sea.

- David dances before the Lord in 2 Samuel 6:14 when he brings the Ark of the Covenant to Jerusalem.

- Psalms 149:3 and 150:4 tell of ways we praise God's name with dancing and playing instruments.

When we use prayer postures and teach them to our children, we model how people have prayed for thousands of years.

- In 1 Kings 8:54, Solomon kneels and stretches out his hands toward heaven.

- Psalm 95:6 calls us to worship, bow down, and kneel.

- Daniel 6:10 describes how Daniel knelt three times a day to pray to God.

- After the Last Supper when Jesus was in the garden, he knelt down and prayed to his Father (Luke 22:41).

Quick! Name a recent time when you worshipped or prayed while moving your body. If you come up short, your little one may be a great source of inspiration to you. Whether through dance or prayer or other activities, you can pair movement with words or songs of faith—or use the movement itself as an expression of your beliefs.

Little Steps

- **Your Whole Body, Parenting:** Print out a picture of yourself or draw an outline and label all the parts of your body that you use to parent your child. Your fingers get tangles out of hair. Your knees bend so you can crouch to see your child face to face. Your lap and arms provide a secure snuggle space.

- **Speaking Scripture:** Learn a few Bible verses by heart that describe ways that our bodies are part of how we express our faithfulness to God. Recite them to your child.

- **Massage:** Infant massage has demonstrated benefits for preterm infants, but all babies can benefit from gentle, repeated touch—and so can their parents. Take a class or look up some techniques online to start this practice with your child. As your little one grows older, look for ways they enjoy your gentle, repeated touch, and find ways they can gently touch you too. Introduce words of prayer and blessing during this time.

- **Breathe Together:** Learn some breathing techniques you can teach your little one. (Note: Teach while kids are calm, not in the middle of a tense situation.) Many options are available that use images and words to help young children visualize how to control their breath, such as imagining they are hibernating bears breathing slowly and deeply, or that they are blowing out candles or blowing feathers or bubbles.

- **Get Moving:** As parents, you can try out ways to model faith for your child through your movements together. See which ones your child enjoys and that you feel comfortable doing. For example, create a dance party playlist of songs based on Bible verses and other words of faith. When your kids move and dance while hearing words about God, love, and the people of the Bible, they make connections between joyful freedom and living the Christian faith.

8

Routines and Rituals: Do It, Then Do It Again

Have you ever heard the saying "Steal from the best and make up the rest"? The faithful parenting ritual I've practiced with my children since they were infants wasn't my idea. Instead, I learned it from a dear friend, Tera, who started her parenting journey a few years before mine and now has three kids. When I was pregnant with our first child, Tera told me of this ritual that she and her spouse practiced with their children, and I thought, "I am totally going to use that."

Here it is: Every night before my children go to sleep, I make the sign of the cross on their forehead as I say, "Paavo, child of God" and "Svea, child of God."

This blessing started when they were babies. I think it was one of the ways they learned their own names—and the name of God.

It continued into their toddler years and introduced an important faith concept to them: "You are our child, but you are also God's child."

By their preschool years, they were learning that the mark of the cross had meaning. It was the shape they saw their pastor trace on the forehead of the newly baptized. It was also the place of Jesus' death. They saw cross imagery throughout our church's worship space, in their Sunday school resources, and embedded in artwork in our home.

By the time my children were in elementary school, they had learned quite a trick. If I forgot to give them their bless-ing, they would stay up reading. When I noticed the light on and walked into their room with the question "Why are you still up?" they would respond, "You didn't give me my cross yet." They had learned that the practice of blessing marked the end of each day.

I'm not sure if and when this practice will stop. I'll keep crossing as long as they keep holding still for it.

Our Christian faith is rooted in our identity as children of God and our belief in Jesus Christ. What do you want your child to learn, remember, and do throughout their faith journey so that they develop a deep and lifelong faith in Jesus Christ? This chapter explores the profound and lasting ways in which you as a parent can introduce simple routines and rituals—a ritual may be described as a routine to which special meaning is added—to support your child in taking little steps toward big faith. This section of the book is centered on an Old Testament text that highlights the importance of the words we speak as parents. This passage is ancient, but because of what contemporary research has revealed about how we learn and retain information, the verses give us new ways to help our children remember and grow in faith. First, read the passage in full. Then we'll dig into it one section at a time.

> *"Hear, O Israel: The LORD is our God, the LORD alone. You shall love the LORD your God with all your heart, and with all your soul, and with all your might. Keep these words that I am commanding you today in your heart. Recite them to your children and talk about them when you are at home and when you are away, when you lie down and when you rise. Bind them as a sign on your hand, fix them as an emblem on your forehead, and write them on the doorposts of your house and on your gates."*
> *—Deuteronomy 6:4-9*

Moses spoke these words of God to the Israelites during one of his sermons as they prepared to enter the promised land. At this point they had been traveling through the wilderness for decades. The babies born on this trip had grown up and had their own children.

These words form a central part of Jewish faith practices, called the *Shema*. This is the oldest daily prayer in Judaism and has been said each morning and night by people of the Jewish faith for thousands of years. As Christians, we recognize within this passage the first part of what Jesus proclaimed as the greatest commandment.

But modern research on learning and memory can provide another lens through which to view this text: God is instructing believers to pass on faith according to how the brain learns best.

Listen Up!

"Hear, O Israel: The LORD is our God, the LORD alone."

These first words are what give the passage its name—the Shema. That's because the word *hear* in Hebrew is "shema." (A modern equivalent might be the command "Listen up!") Once Moses had the people's attention, he proclaimed the statements that follow, which formed the basis of their belief. The words affirm that there is only one God. This was a bold and unique assertion because people worshipped many gods during the time of Moses, as well as thousands of years later during the time of Jesus. Thus, these words are foundational to both the Jewish and the Christian faith.

Then Moses moves on to a command: **"You shall love the LORD your God with all your heart, and with all your soul, and with all your might."**

This command is echoed elsewhere in Deuteronomy, and in Joshua. Later, Jesus will speak these words when he is asked to name the greatest commandment. Jesus also expands on Moses' words by adding "Love your neighbor as yourself" to the commandment. As Christian parents, we, too, want our children to hear and believe these words.

To reiterate: Moses proclaimed the core of the Israelites' faith—that the Lord is God, and we are to love our God wholly. *This is what we believe*, he tells the Israelites. But he doesn't stop there. God has made it clear that it's important to remember these words, so Moses tells the people, **"Keep these words that I am commanding you today in your heart."**

But how? How can God's people keep God's commands in their heart, day after day? Moses has a suggestion here too—and it's in line with the learning and memorization methods psychologists recommend today!

Pass It Along

"Recite [these words] to your children . . ."

Moses began by telling the Israelites to "listen up" to the words that would provide the foundation of their faith in God. Now he moves on to tell them how to *live* these words and pass them along to future generations. Moses is giving parenting advice! He tells parents and grandparents that these words of faith are so important that they need to be spoken to children as well as to grown-ups. He uses the term *recite*, but other biblical translations use *impress*, *repeat*, and *teach diligently*. It is not enough to say words about God only once to the children.

Moses instructs his listeners to say them over and over. This idea of repetition has received a lot of support in brain science.

Research on human memory has shown that our brains are designed to remember content that is repeated, or rehearsed. (Think back to how you memorized the states and capitals in school, or to a time when you had to remember a phone number because you couldn't write it down.) The purpose of rehearsing something is to encode that content so deeply that it is stored in our long-term memory and recalled when we need it. When the Israelites are told to repeat Moses' words to their children, the stage is set for this rehearsal to happen so that children will remember.

Researchers who study memory in early childhood have learned that infants can comprehend common words well before their first birthday—in fact, as early as six months! Other studies demonstrate that toddlers are significantly more likely to remember newly introduced words when they have cues and support about what these words mean. Our toddlers need us to explain and describe new words; when we do, they remember!

Overall, researchers have found that memory retention for words improves steadily during the first two years of life. As children hear more words, understand more words, and speak more words, they remember more words! But how do these findings connect to God's command in the Bible passage to "recite them to your children"? In short, it is a powerful cue to us as parents to begin speaking words of faith right from the start so that our children never remember a time when they were not hearing about God's love for them and their love for God. It's never too early to have a conversation with your parenting partner and others who care for your child about the words of faith you want your child to hear over and over.

If we know that we need to keep repeating words of faith to—and impressing them on—our children, where, when, and how can we do so? The remaining verses of the passage address these questions. What's more, the practices mentioned in these verses are supported by research findings that establish them as well-documented techniques to aid remembering.

". . . and talk about them when you are at home and when you are away, . . ."

Moses tells the Hebrew people to speak words of faith in multiple places—not just in the security and familiarity of the home. (At the time, their homes would have been the tents they used during their forty years of wandering in the wilderness.) Notice, too, that the command is to "talk about them," not just recite the commandments verbatim; we help our children learn not just by repeating the same words over and over, but by talking about what these words *mean*. But let's focus in particular on the words "when you are at home and when you are away." This command establishes that faith conversations should happen in various locations. How amazing that Moses' words from millennia ago point us to recent findings that *what* we learn is influenced by *where* we learn it.

Researchers use the term *context-dependent learning* to describe the ways our learning can be shaped by where we learn something. Results across all age groups have shown that memory about something we've learned is more accurate when we are asked to recall it in the same context as where we learned it, as opposed to somewhere else.

We can apply findings from studies of context-dependent learning in children as young as infants and toddlers because researchers have specifically studied word learning in children of this age. They've found that beginning in infancy, children retain words longer if they are asked to remember them in the same context where they learned them.

At the same time, however, we see that this Bible verse doesn't tell us to teach the words of faith *only* at home and then remember them *only* at home. It tells us to talk about them at home *and* away from home. That is, we are commanded to speak about God to our children in multiple places. When we learn something across multiple contexts, or settings, our learning often goes deeper and lasts even longer than if we just learn it—and recall it—in one place.

This brings us back to the verse: "Talk about [these words] when you are at home and when you are away." Here we have God calling the ancient Israelites—and us—to talk about God in many places, not just our homes. Living our lives as followers of Jesus means living out our faith no matter where we are and modeling that to our children. When we consider that young brains are likely to remember conversations more deeply when we have them in multiple places, that is all the more reason to use faith language both at home and on the go.

If you're wondering *how* you can use faith language away from home, start by looking at the books you have tucked in your diaper bag. You already know that reading to your child is a great way to build literacy (and fill time at the pediatrician's waiting room or an older sibling's soccer practice). When you include children's Bibles and faith-based storybooks in your stash, you reinforce the idea that your family uses faith language in all kinds of places, not just amid the familiarity of home. Another idea to try is saying mealtime prayers *everywhere*, not just around your kitchen table. Whether you're at a restaurant, eating at a friend's home, or grabbing a quick bite while in the car, saying a prayer of thanks before eating teaches your child a faith practice they can use for a lifetime.

". . . when you lie down and when you rise."

Moses also instructs the Hebrews to remember and speak these words about God at two specific times of day. The result is a blueprint for morning and evening prayer. At the end of the day, before settling into sleep, remember that the Lord is your God. The next morning, before getting started with the day, remember that the Lord is your God. Whatever the day may hold, it is marked at beginning and end with a proclamation about who God is and how we are called to love God.

Once again, Moses' suggestion of how and when to speak words of faith has a lot of backing in early childhood research. If you're a new parent, I'm sure you can attest to the major role sleep (or lack of it!) plays in the early years of a child's life. Adults generally "lie down and . . . rise" once every twenty-four hours—but we know that with young children who nap, these bookends around sleep happen a few times a day, not just in the mornings and evenings.

Researchers who study human memory have looked at how sleep affects learning with infants, toddlers, and preschoolers. In some of these studies, the researchers teach something new to a group of children, then have half of them take a nap and half of them stay awake. (This sounds like the most difficult research project in the world to me!) These studies have found that the children who nap after learning remember more than those who stay awake. (Such findings on sleep and learning are true with adults as well: when we "sleep on it" after learning something new, we give our brains a way to consolidate the information so we recall it better later.) Our brains are designed so that sleep helps us—whatever our age— to remember. And incredibly, here we have this passage in Scripture where God tells the Israelites, "Speak these words to your children before they sleep"!

All parents have stories to tell about their children's sleep habits. Many of us focus on the challenges of getting our little ones to sleep, what happens when they don't stay asleep all night, or how we are facing the challenges of being sleep deprived. But think beyond the trials and tribulations of children and sleep for a moment, to the *opportunity* of sleep. Every day, you have multiple chances to speak words of faith with your little ones as they get ready to go down for a nap, when they're going to bed at night, and during the groggy, vulnerable transition as they wake up from sleep.

These times of moving in and out of sleep are holy times, sacred moments of drowsing off and waking. As you soak in these lovely moments with your little one, try introducing faith language in short and simple ways. Your voice may be the last thing your child hears before slipping into sleep for a nap or for the night. As their brain prepares to consolidate the day's experiences, you could gently speak or sing words of faith, as a prayer or a blessing or a Bible verse, to reinforce God's love and care for them. When you greet your child with words of faith as they begin to stir in the morning or post-nap, you reinforce your family's beliefs at a time when your child feels fresh and rested. Words of faith spoken during these transition times are more likely to be remembered, especially if they are paired with loving touch. (See the next section.)

"Bind them as a sign on your hand, fix them as an emblem on your forehead, . . ."

This part of the passage may not be as easy to understand as Moses' previous suggestions to repeat words of faith, to speak them in multiple locations, and to speak them when we go to sleep and when we wake. Here, Moses is referring to the Jewish practice of literally attaching words from the Hebrew scriptures to one's hands and forehead in small leather boxes called *tefillin* or *phylacteries*. Though this practice is foreign to many of us, the idea of actually tying God's words to our bodies can inspire us to accompany words of faith with physical actions and loving, caring touch, and thus to make our words even more memorable. (See also the previous chapter on whole-body parenting.)

We know that touch plays an important role in healthy development throughout early childhood. Safe, loving touch offers comfort and security to infants, who learn about their caregivers through physical interactions paired with words, facial expressions, and other kinds of social input. As babies enter toddler years, meaningful learning experiences typically happen through more than just one of the senses. When a child experiences multisensory input—especially when their sense of touch is paired with what they see or hear—their learning goes deeper and lasts longer.

You can combine your words of faith with many forms of physical contact. For example, you can hold your child's hands or clasp them into a prayer posture as you say or sing a prayer. You can cradle their head or make the sign of the cross on their forehead as you speak words of blessing, like I do in my son and daughter's bedtime ritual. Infant baptisms and baby dedications are powerful times of spoken words and physical touch. Whether your faith tradition splashes the waters of baptism or anoints with oil during a dedication, using words and touch together echoes God's command to tie the words of faith as symbols on our hands and bind them to our foreheads.

". . . and write them on the doorposts of your house and on your gates."

This part of the verse points to another practice in the Jewish faith: that of literally attaching the words of the Shema to structures in the home in a *mezuzah*. This small case holds a scroll with the words of Moses from this passage; it is affixed on doorframes in some Jewish homes yet today. The presence of these words, in ancient times and now, would bless the home and communicate to anyone who passed through the gates that the family followed God.

The developmental principle at work in this part of the passage is how visual cues assist with learning and meaning-making. Research on visual cues supporting memory begins with babies. When we can show babies, toddlers, or preschoolers something that corresponds to our words, they will be more likely to remember what we are saying—and what we are showing them. This also ties into how our memory tends to be more effective when we receive information through more than one of our senses.

But this part of the passage is about more than just visual cues. The idea of writing God's words on our doorframes leads us to think about our homes, and what they tell the world about what we value and believe. The art on our walls, the papers on our fridge, and the books on our shelves all send messages to the people who live in and pass through our home: *This is what's important to us. This is how we express ourselves. This is what we want to learn more about.*

If someone walked into your home right now, how would they know you were a Christian family? How could you surround your child with pictures, symbols, words, and other signs of your family's beliefs so they never remember a time when your home didn't have on its walls a way to say, "This is who we are and what we believe"? Here are a few ideas: Look for Christian symbols to place in your home for your child to notice. (A cross, a dove, and a shell are all good starts.) No matter what nursery theme and décor you've chosen, find room to post a Bible verse, words of blessing, or other messages you can repeat to your child. Start simple by focusing on your entryway, choosing a word or symbol of faith to greet your family and visitors each time someone passes through the doorway.

As a parent, you have already seen the power of routines. The stability they offer your child (and you) produces predictability in the face of the inevitable changes and disruptions of family life. But even when a routine seems to be working well, you may need to change it to accommodate the fact that your child is developing quickly and needs a new way.

You can add meaning to some of the routines we've discussed to create rituals that support the beginnings of your child's faith. The words of the Shema can inspire everyday actions supported by decades of research on human memory. You may find yourself inspired to explore other rituals as well.

Little Steps

- Write the words of Deuteronomy 6:4-9 someplace where you will see them regularly.

- Think of all the places your family goes in your community during a typical week. Choose a new place for speaking words of faith with your little one.

- Try saying a prayer, reading a Bible story, or speaking a blessing to your child during your typical naptime or bedtime routine. Pair these with touch: making the sign of the cross on your child, holding hands together for prayer, taking time for a loving hug.

- Look around your home to see what words, pictures, or symbols you can find—or create some out of basic craft supplies—and display them where both children and adults will see them regularly.

9

Community:
Why Churches
Are Rich Contexts

When my son, Paavo, was two and our godson, Erik, was a baby, our families were sitting near each other during a worship service. As she did each Sunday, Pastor Pam called all the children to the front of the sanctuary for the children's message. Paavo began walking toward the other children who were assembling up front, but then noticed that baby Erik, nestled comfortably in his car seat, wasn't being picked up by a parent to join the other children. He stopped in his tracks, turned to Erik's dad, and asked, "Joe, is Erik coming to the children's message?" (Yes, Paavo was that verbal at two!) Joe responded, "No, Paavo, he'll stay here."

A second time Paavo inquired, "Joe, is Erik coming to the children's message?"

"No, Paavo, we'll stay here," Joe repeated.

The third time my son asked, Joe realized it wasn't really a question. Instead, Paavo was inviting Erik to join him by pressuring his father to carry him there since Erik wasn't yet able to get there on his own. Erik went up for the children's message for the first time that day (and many, many more times that followed).

We treasure this story in our family as a vibrant example of our young child's ability to show the power of invitation and welcome because it was so deeply ingrained in his community. He simply could not conceive of a reason that would prevent every baby from joining the older kids to hear the message.

Ask any new parent what a perfect Sunday morning looks like. "Going to church" is likely *not* the top answer. You're more likely to hear responses like sleeping in, eating a late breakfast, playing together, or a range of other activities that don't have the urgency and underlying timetables and schedules that make up the rest of the week. I've heard some parents describe these relaxed Sunday mornings as worshipping at St. Mattress or the Church of the Holy Comforter.

Indeed, eating meals with family, playing with parents, and enjoying a relaxed home environment are all activities that support healthy brain development in young children. But in this chapter, I'll try to persuade you that waking up in time to get everyone dressed, fed, and loaded into the car for church provides your little one with some unique and remarkable opportunities that stimulate their brains and help build a foundation of faith that begins one little step at a time.

Before I explain the many ways that churches can provide a rich neurological context for early development, I want to let you know I sympathize with how new parents struggle to get to church regularly. Churchgoing *is* a difficult activity for families with babies and toddlers, involving the complicated wrangling of snacks, diaper bags, nap schedules, and maybe the kind of clothing that kids would rather not be wearing. Additionally, many churches would confess that they struggle to know what to do when it comes to welcoming families with young children. It may surprise you to know that your family is a desirable demographic for church leaders. But it's true! What church doesn't want to describe itself as "the one with lots of families with young children"? Even so, many congregations are not yet examining how they may need to revise their programs, plan worship services differently, and redesign parts of their physical space to make church more welcoming to the young families they'd like to see in their pews.

In my years of studying and teaching about early childhood ministry, I've listened to many parents of young children tell their stories of why they are (or aren't) regularly attending church. While these stories have contained many variations, I've noticed some recurring themes. I've been able to identify four basic stories about parents' relationship with church: The Returner, The Switcher, The New Joiner, and The Good Fit. Read below and see which story most closely fits your family's situation. For each one, I've offered guidelines to keep in mind as you explore what your relationship with a church community will be during your child's early years.

The Returner

"I grew up going to church but haven't gone as an adult.
Now that I've had a child, I feel a tug to go back."

If you're a Returner, my main piece of advice is to be open to a church with ministries and worship services that fit your life *now*, instead of being set on finding again what you remember from your growing-up years. It's okay to let go of expectations from your childhood and youth. Now that you're an adult, you have different needs and expectations.

So, what should you look for in a church? If possible, look for a church with other young families, and particularly other Returners. Their stories may mirror yours, and you may find that this micro-community within the church provides you with the kind of support you are looking for.

Above all, be patient with yourself and your family. You are making a bold move to reenter a church community when so much in your life is changing. Tend to your own faith too. Church isn't just a place where you get reinforcement as you teach your child values and morals to live by. Church can be a community that supports your needs as a parent, including the ways in which your own faith is deepening.

The Switcher

*"The church I belonged to before having kids
doesn't seem like a fit right now."*

Switchers sometimes feel guilty about feeling dissatisfied with a faith community that was a good church home before they had kids. That guilt is unhelpful—and unnecessary. You don't have to feel obligated to stay at a church community that's no longer meeting your needs now that you've undertaken the enormous job of parenting a child in the Christian faith. If your current faith community isn't adjusting to your needs now, they probably aren't going to change their tune unless there are major forces in the church clamoring for change. (Since parents of young children often don't have the time or energy to be that force, it's not likely to happen.)

Since you're already a member of a church, you have an idea of what you like and what you don't when it comes to worship style, ministry programs, hospitality, and a church's overall "personality." If you decide to look for a new church, try making a list first before you visit. What are you looking for in a pastor's leadership style? In what types of ministries and mission outreach do you hope a church invests its time and dollars? How does a worship service need to fuel you and connect you to others in the church family? What practices and characteristics will assure you that a church welcomes and values young children? Answering these questions may feel like a weird variation on speed dating, but if you know what your expectations are, it will be easier to weed out the bad fits and find the place that will welcome you and your little one. After all, you wouldn't return to a restaurant that didn't have a high chair, kids' menus, or a waitstaff that was friendly to your child. It's okay to be direct (and selective) with what you need from a church community too.

If anyone asks you why you made a switch (and tries to guilt you for it), you can let them know you were looking for a place that would welcome your whole family and support you in raising your child in faith. Isn't it great that the immediacy and enormity of parenting little ones gives you permission to state the truth and get on with things?

The New Joiner

"I've never belonged to a church before, but now that I have a child, I'm wondering if I should check one out."

Your tug to join a faith community right now might be prompted by several factors. Maybe you have friends with kids who are part of a church community and there is something compelling about the stories they tell. Maybe you've witnessed how a faith community helped a friend in deep need. Maybe you've identified in the past as "spiritual but not religious" but feel like parenting is calling you to a more formal and communal expression of your spiritual self. Whatever the reason, open yourself up to the possibility that you and your child may be welcomed by a faith community in beautiful and unexpected ways.

As the busy parent of a little one, you likely don't feel you have time for in-depth exploration of the theological positions of the many church denominations that exist. (If you are interested in learning more about different Christian denominations, I recommend the book *Honoring Our Neighbor's Faith*, by Robert Buckley Farlee, for a good overview.) But do take time to talk with your friends, look at the churches in your immediate community, and do some online sleuthing. If a church you visit seems friendly but you don't agree with some of its major beliefs, it may not be the best fit. Similarly, if a church that looks right on paper or online treats you with dismal hospitality, it is fine to cross it off the list.

Trust your gut. As a new parent, you're developing a finely tuned sense for what seems right for you and your child. Choosing to connect with a church community means opening yourself up to new experiences during this vulnerable time of life and entrusting others to build relationships with you and your child.

The Good Fit

> *"The church I belong to now is an awesome place for me and my family."*

If you're in a Good Fit church, consider yourself lucky! You're blessed to be part of a faith community that adapts to your growing family's new needs. If you find yourself blessed in this way, the best thing to do is pass some of that blessing on to others! Tell your story to other parents who may be searching for a church home. They are much more likely to trust you than a church website or recommendations from childless people or those whose kids are much older. You may turn into one of those people who has a whole trail of folks who trace their initial connection with your church to you!

Affirm your church's leadership when you have the time and energy to do so. I'm not talking about handwritten thank-you notes here—just a quick connection sometime when you see them at church, or maybe a quick message on social media. Your voice matters—a lot—to your church leaders.

Even if you love your Good Fit church, don't feel like you have to shoulder lots of volunteer responsibilities at this time. Yes, your church may be recruiting people to deliver meals, decorate the church for Christmas, or help with affordable housing initiatives. Families with young children are typically not in the season of life where they can reliably show up and help with these kinds of events, even if free child care is provided. Planning a time of family service can be simpler than this—donating to a local crisis nursery, making cards from your child's art to send to others in the congregation, helping pick up toys in the nursery after your little one has spent some time there, and so on.

No matter which identity you claim, the process of finding a church where you and your family feel comfortable is a journey of belonging. *Belong* is the verb I use when talking about a family's relationship to a church community. While the term *membership* is used to convey a more formal relationship (taking a class, officially joining the congregation, committing to regular financial support), I believe the term *belonging* is a more powerful one to describe our feelings about a church community. When we belong (even if we aren't formal members), we know we are welcome. We feel comfortable and at home when we are in a place we belong. For new parents, this sense of belonging is especially pronounced when we consider how our children respond to being in the church building. (The inverse is true as well, unfortunately. Many people describe a situation in which they are members at a church but do not feel any sense of belonging. Those parents are welcome to become Switchers.)

At its best, a church community can be a vibrant context for early development because of the distinctive experiences it offers to young children and their families. Using the topics from the other chapters in this book, consider how your child's development is enhanced each time you participate in your faith community such that their brain can form and reinforce synaptic pathways as neurons fire together and wire together in healthy, faith-nourishing ways.

Attachment Relationships

Children form attachment relationships with a small number of people who provide them with regular, sensitive care. The security (or lack thereof) of these attachment relationships sets the stage for the relationships a child forms with others. You may remember how, in chapter 3, I described the ways that your love for your child provides a model of God's love for them. In a church setting, your child will have a chance to form a wider range of relationships that mirror the loving community that the Bible refers to as the body of Christ.

A pastor is "shepherd" to the "sheep" of the congregation, modeling wisdom, guidance, and care for all in the community. Intergenerational relationships flourish when other adults get to know the young children in a church and are invited by parents to connect with their family. You may witness them cooing over newborns, asking about your baby's new developments each time they see you, and delighting in the exuberance of the toddlers in their midst. Older adults without grandchildren (or whose grandkids live far away) may take special interest in little ones.

Your child may also develop deeper positive relationships with some of the people at church. Skilled church nursery caregivers will recognize children, call them by name, and learn their toy, book, and snack preferences. Because little ones are drawn to "big kids" and can learn much from them, they may form relationships with youth who serve as nursery helpers, or with an older child who is just a familiar and fun person to see at church.

Witnessing and being a part of healthy relationships at church helps children know that a church family is a place where they can feel secure, known, and loved.

Language

As you read in chapter 4, your child's language development can be supported right from the start. Speaking to your child many times a day using a wide vocabulary can include words of faith to introduce your child to who and whose they are.

Hopefully, you and your child are greeted with warm, welcoming language each time you walk through a church's doors. Encounters with fellow churchgoers may include quick greetings in the hallways, laughter-filled conversations over snacks and coffee, or kind exchanges in the bathroom while you change a diaper or help your child use the potty. Your church may have nursery care where you drop off your little one before heading to the adult-only class or discussion group. During their time there, your child may hear Bible stories, prayers, and loving words spoken by the nursery caregivers.

But the central place where your child will benefit from hearing the language of faith is during worship with the community. When a young child is present with other worshippers, they are bathed in language about who God is, what God has done, and why we gather to worship God. They hear lessons read from Scripture. They become familiar with the rhythmic speech and emotional tone of public prayer. They learn the cadence, pace, and sound of the pastor during the sermon and other times of worship leadership. And they witness a sound like no other when they hear the people around them speaking the same words together during greetings, sendings, and prayers.

This exposure to the language of faith ensures that your child will never remember a time when they didn't hear familiar people of many ages expressing their faith together and proclaiming God's word.

Literacy

Closely linked to language development is early literacy, described in chapter 5. With so many children's Bibles and faith-based books available for your child, supporting their understanding of print can happen at the same time as introducing stories of faith.

Your church can also support your child's spiritual development through its emphasis on literacy. The church is a literate place where printed words matter, which means that church is a setting that supports emergent literacy in young children. Children may notice letters and words in the hallways, on signs, and even painted on the walls. Words of faith surround your family during worship, whether in artwork, on paper, or projected onto a screen. And the church roots itself in God's

Word, which means Bible stories are read to all who gather, whether they are younger than one or older than one hundred!

While many settings can support early print literacy—homes, day cares, preschools, libraries—church is the place where *biblical* literacy can be supported from the start so that young children never remember a time when they didn't see God's Word represented in text and hear it read aloud.

Music

As chapter 6 explains, we know that young children are attentive and skilled listeners of music. Starting in infancy, children can comprehend musical structure, tempo, and emotions and use their bodies to respond to music. Your family may listen to customized playlists at home, at day care and preschool, in the car, and while at others' homes. Additionally, music often plays in the background at retail stores and community spaces. Most of this music is prerecorded and treated as a way to create a mood or make shoppers stay in the store longer.

Music at church can sound different—a lot different—than at these other places. Whether children are in nursery care, a parent-child class, or Sunday school, their leaders often choose faith-based music that is composed and produced specifically for young participants. They can sing the songs with your child, pointing out the lyrics that teach Bible stories, Bible verses, and other faith concepts. They can lead young children and their parents in music and movement, making the connection between Bible-time people who danced to praise God and the little ones in a nursery or classroom who are, themselves, learning to praise God through dance.

In addition to the music heard in church settings designed just for little ones, your child experiences unique acoustic events when they attend worship with you and the entire church family. Here are a few ways your young child may hear music during a service:

- A musician guides back-and-forth singing with the congregation. Your child witnesses how one strong, pleasing male or female voice takes turns singing with everyone else in the group.

- A small group or choir sings a song accompanied by a piano, organ, or other instruments. The child hears voices singing melodies and harmonies and watches the animated faces of those who produce this live music.

- A vocalist sings a solo during the service, and a child sees how one person produces loud sounds and soft sounds, high notes and low notes, all in real time, with one or more other people accompanying with instruments.

- Whether the church worships with the acoustic sounds of a pipe organ and hand bells or the electronic sounds of bass guitars and keyboards, a child is surrounded by music that hums and vibrates, swells and diminishes. All the while, they can see the people who are making this music (and may even catch a musician's eye).

- The "surround sound" of congregational singing provides sonic stimulation unlike anything else!

These varied musical experiences can make an impression—literally—on the minds of the youngest ones in our midst. Long before they can read music or play an instrument, they learn that singing and playing instruments are what the community of believers does together.

Whole-Body Parenting

In chapter 7, I proposed the idea of whole-body parenting, using examples about breath, touch, and movement to show how parenting is an embodied call that can lead us to actions expressing our faith in ways that go deeper than our words.

Parenting a child while at church is a whole-body experience, with lots of lifting and carrying, rocking and swaying. You may lose track of how many times you scoop up your little one and set them back down, whether placing your baby in the car seat or letting your toddler explore the surrounding spaces. You may hold them close as you sing or pray. You may give them a little kiss after you receive Holy Communion so that they smell the bread and wine you've just received.

Belonging to a church community isn't a drop-my-child-off kind of experience. Instead, your child is close by and you have opportunities each time you are there to breathe together, move together, and make physical connections that may be different than at any other time of the week.

When your child learns that church is a place where they are embraced by members of the whole body of Christ who love them dearly, they can make deeper associations with God's tender love for them.

Routines and Rituals

Over the centuries, the Christian church has developed routines and rituals that are still practiced by millions of believers today. Many of these routines and rituals are rooted in the words of the Shema, and as chapter 8 explored, these early practices are backed up by some very current research on human learning and memory.

Church can be a place for you to learn new routines and rituals that will grow as your child does. Think about the physical routines your child sees during worship, like a pastor raising arms to welcome people or making the sign of the cross. You may also kneel, stand, and sit during worship. Different churches have different traditions for prayer postures as well. Some of these worship rituals can be imitated by young children, so you can encourage their direct participation during worship and at other times too. Raise your arms, fold your hands, kneel, cross yourself, and see what other movements from worship you can weave into your daily lives together.

Many parents have noted that their toddler or preschooler plays "church" at home, organizing dolls, stuffed animals, and other figures to sit together for readings, sermons, and baptisms.

(Expect this play to include other events and characters too.) This can be a great opportunity to witness what little ones are noticing about church, so watch carefully!

During his ministry, Jesus invited little children to come to him, and told his followers, "Do not stop them; for it is to such as these that the kingdom of God belongs" (Matthew 19:14). Having children present in a church community is a regular reminder of just who Jesus was talking about. When older members of the community can witness children's vulnerability, dependence, and capacity to love deeply, this window into the kingdom of God becomes clearer.

Little Steps

- Whether you're just starting to look for a church community or are established at a place of worship, commit to attending as often as you can. Maybe that means shifting around some weekend activities, packing the diaper bag and setting out everyone's clothes the night before, or adjusting other family events and routines so that going to worship at church becomes part of what your family does regularly.

- Take a picture of your church building, print it, and post it someplace where your family can see it every day. Point to it, talk about church, and let your child know it is an important place for your family.

- Pray for people and ministries in your church out loud so your child hears that your prayers include the people and things from church.

- If possible, play recordings from your church's worship services so your child can hear this music at home and on the go. Soon they may be singing and dancing along.

- Open yourself up to the idea that church is a place where you can share your child with the community. Some people in your church community may no longer have family members who are little ones, or their young relatives may live hundreds of miles away. Your child's presence is a gift to the community, and your church community can offer you support in ways you hadn't imagined.

10

Now What?
How to Live It

My daughter, Svea, has been a ballet student since age three. The studio plans two recitals each year for family members and friends. Each program progresses in the same way—the youngest dancers begin the show, and the subsequent groups become more and more advanced until the oldest, most skilled dancers perform the last piece.

At her very first recital, Svea sat excitedly in her group of three-year-olds, a gaggle of little dancers wearing pink leotards and simple hairpieces. As the youngest group, they began the program. We delighted in seeing her and the other students in action as they followed their teacher's lead in leaping, spinning, walking, extending arms, pointing toes, and more. Except, of course, none of these actions happened in perfect synchrony. A few dancers followed the teacher's every move, some were on a five-second delay, and a few seemed a little dazed by the whole prospect of being in front of a hundred people watching them so closely. Still, the applause was vigorous and heartfelt when the piece ended.

These dancers had done what they could at their level to dance what they knew. Once they'd completed their piece, they sat back down in their assigned seats. Then another amazing thing happened: I watched these girls become transfixed by the older dancers who performed song after song. They recognized their connection to these kids. If I keep dancing, they may have wondered, could I be the one doing that someday?

As we've watched our dancing daughter grow, we've seen Svea and her peers progress through the program so that now they perform near the end. Their movements are lovely and fluid, their coordination nearly seamless. We aren't just applauding them for effort (and cuteness) anymore—we're celebrating the beautiful artistic result of their hard work and commitment.

Svea's earliest years as a dancer provided a foundation of skills that she's built on with each new year of study. Part of her dance education has been to watch the older children show her what is to come if she continues to dance. This is an important part of the process for parents as well. No child is born knowing how to dance. They need their parents to support them by signing them up and showing up so they can learn and grow.

For your child, the "dance" has begun. They are already learning who they are, who you are, and what kind of world surrounds them. You are providing a foundation for them to learn the first basic steps of language, literacy, and music. But unlike those parents at the dance recital who sit in the audience watching their children dance, noting how they change and grow from year to year, you are part of the dance too. So how can you support your child (and yourself) in growing and learning more about the dance of faith? Following are seven suggestions of some next little steps you can take to help your child grow toward big faith.

Dare to state your hopes for your child's faith.

Your child is growing up in a culture that views religion differently than during your own early childhood and that of your parents, grandparents, and previous generations. Growing up in the Christian faith, belonging to a faith community, and then raising one's children in the same faith was a pattern many families followed for generations. In some families, parents didn't speak often to the children about these expectations, their own personal beliefs, or their hopes for how their children would grow into mature Christian adulthood. Instead, many families simply showed up, participated in religious activities, and assumed their children would do the same.

Well, times have changed! Various studies of religious life in the United States report that among younger generations, fewer people identify with religious groups than did so in previous generations. Among those who do, fewer attend worship services, pray, and read scripture compared to older generations. A friend summed things up when he told me, "Dawn, we used to be the home team. Now we're the visiting one." Rather than viewing this shift in religious life as a threat or an insurmountable challenge, we can see it as an exciting and vibrant time to raise our children in the Christian faith. When families choose to share their faith stories and affirm their beliefs more boldly, their children grow up knowing what is most important.

Align with your parenting partner and agree to share the work.

Aligning does not have to mean holding identical beliefs or expressing your faith in exactly the same ways. But your child can benefit by seeing both of you prioritizing their faith formation and encouraging their little steps. In many households, this facet of parenting is not shared equally between spouses. Recent research on religious life in the United States has shown that mothers typically bear more of the duties of passing down faith than do fathers. Whether bringing their children to church, talking about their faith, or praying and reading scripture, mothers are more likely to be the ones engaged in active faith practices. Dads can claim a more active role in their child's faith formation through carrying out the same faith-based routines and rituals as mothers do, or by developing their own unique ones.

What might that look like?

- Both of you say simple prayers with your child.

- You each have a unique way to bless your child.

- Reading from story Bibles and faith-based books isn't just a Mom thing or a Dad thing—it's a family thing that you prioritize during storytimes.

- Both of you are comfortable sharing simple phrases about why faith is important in your family.

- You share words of faith during the routines and rituals described in chapter 8—when you are at home and away, and when your child lies down and rises.

- Together you manage the work of getting the family to church on Sunday morning (or whenever you worship); it's not solely up to Mom!

If you do not have a spouse or parenting partner at this time, maybe you can identify a close friend whose Christian faith inspires you. How could you invite this person into a deeper relationship with your family to help influence your child's faith formation?

Keep learning about your child's brain (and your own).

Researchers in the fields of neuroscience, psychology, linguistics, and education will continue to make amazing discoveries about young brains. You may feel like your parenting focus can't possibly expand past the day-to-day needs of your child, but give yourself some credit—you're finishing the final chapter of this book! Learning just a bit at a time about your child could inspire you to try something new to help them take a little step toward big faith.

- Seek out articles, blogs, or even books on child development topics that are especially interesting to you.

- Ask your pediatrician or other health care professionals for recommendations on how to learn more about your child's brain development and how to support it through enriching experiences, good nutrition, and other healthy parenting practices.

- If you have friends on social media who seem well read on brain development, ask them what they are reading or watching to learn more about this topic.

- When you learn something new about your child's development, wonder about how you might be able to apply that insight to their faith formation.

- Keep in mind that *your* brain is changing too. Experiences can change the human brain at any age, even when you feel sleep deprived! How might your own ways of thinking, feeling, and acting be affected by what you learn about your mind?

Find opportunities that support age-appropriate faith experiences for your child.

Chapter 9 presented a rationale for finding a church community because of the loving and neurologically rich context it provides, especially during worship. A church family can provide strong, caring relationships for you and your little one. Additionally, you can look to other occasions and settings for support in raising your child in the Christian faith.

- You may decide to look for child care and preschool at a program hosted in a church. These early childhood centers and schools often have caring Christian leaders who are well qualified to care for and teach your child, and who also view this work as an expression of their faith.

- Check whether a local church sponsors a moms group, a dads group, family faith classes, or other gatherings of parents and children during weekdays or other times apart from its Sunday morning ministries.

- If your faith tradition supports outdoor ministries, you may find that attending family camp begins a tradition of being immersed in God's creation and enjoying time together in unique ways.

- If you can't find an opportunity in your community, could you start one? Maybe a parent support group/play group that starts on your watch could lead to an ongoing program of faith-growth and mutual care for multiple families in your area.

Do a home faith audit.

Have you ever heard of a home energy audit? A representative from the electric or gas company visits your home to point out ways in which you could be more energy efficient. Try looking around your home—at both the objects you see and the routines and rituals you've established—to determine whether there are some simple ways to be more deliberate about faith language and practices with your child.

- What images, symbols, or words greet you at your door?

- What are a few words and phrases that teach simple truths about your relationship with God?

- What prayers do you know by heart for mealtimes, bedtimes, and other times?

- Which Bible storybooks and other faith-based books are in your child's library?

- Does your child have pictures, wall art, and other images that show Bible stories?

Turn to prayer.

Hopefully, the strategies, ideas, and tips I've provided throughout this book will inspire you to think about ways in which child development research can support your child's faith. But nothing can take the place of opening yourself up to God, asking for help, and listening for how God might be at work in your life as a parent. God may give you a nudge or a hint (or a very obvious sign) about part of your parenting life that needs tending, forgiveness, or release. Just as this book has encouraged you to see how little steps of faith can be woven into everyday activities, you may find that praying to God during everyday moments leads to a deeper relationship with God and with those you hold most dear.

Be open to changes in your faith.

Parenting has changed you, right? As parents, we experience changes in our bodies, in our minds, in our relationships, and in our awareness of the world around us. However you described life before parenthood, the description surely changed once that little one was placed in your arms. Why wouldn't you expect the changes to extend to the deepest part of you—the part of you most closely connected to our Creator, Redeemer, and Sustainer of life?

In the busyness of being a parent, you may not always notice when these changes occur. Take a moment to ask yourself, "How has my faith changed since having this child?" Are you more trusting? More loving? More full of wonder? More in need of peace and calm? Tending to your own faith may make you more able to recognize how faith is beginning to form in your young child.

Do you remember Brad, the dad I described in the first chapter of this book? He was raising his daughter to be a University of Tennessee fan, but when it came to her participation in religion, he was going to have her baptized, and then let her decide about faith when she was older.

Brad was in my class in the late 1990s. His daughter is now an adult. I probably won't ever discover what happened to her. She may identify as a "none," someone who claims no religious affiliation. She may use the phrase "spiritual but not religious" to describe her beliefs. Or she may have had powerful experiences that led her to claim her identity as a beloved child of God.

You have tremendous power as the parent of a young child to introduce them to the Christian faith and the welcoming embrace of a Creator whose love you reflect in your own relationship with your child. Go in peace to claim that influence and shape your young child's mind (neural pathways and all) and their earliest experiences, to help them grow into mature Christian adulthood and big faith. God will be with you every little step of the way.

For Further Reading

You made it to the end of the book and you want to go deeper? Fantastic! Here are a few books on child development and faith formation, along with two Bible recommendations.

For more stories:

The Spiritual Life of Children
by Robert Coles

Robert Coles is a child psychiatrist, Harvard professor, and Pulitzer prize-winner, but don't let his credentials overwhelm you. In this book he listens deeply to kids. Coles interviewed hundreds of children around the world from different religious traditions about their understandings of God. The book reports on fascinating exchanges that he had with these 8- through 12-year-olds.

For more doable ideas:

Real Kids, Real Faith:
Practices for Nurturing Children's Spiritual Lives
by Karen-Marie Yust

This pastor and professor has written a wise and helpful book for parents seeking practical ways to support their children's spiritual development. Dr. Yust's work is based in part on a multiyear study on children's spirituality. Throughout the book, she assures parents that their children can engage in activities that cultivate spiritual development in powerful, lasting ways.

For those working in children's ministry:

Will Our Children Have Faith? Third Revised Edition
by John Westerhoff

First released in 1976, for decades this book has challenged and inspired those who work in children's ministry. The title question may sound alarmist, but it adds urgent support to the underlying message—children's faith formation cannot happen only during the time they spend in Sunday school. Instead, children need opportunities for formation all week long. Fortunately, churches can equip parents to do just that.

For assurance that you are doing enough:

The Gardener and the Carpenter:
What the New Science of Child Development Tells Us
About the Relationship between Parents and Children
by Alison Gopnik

Alison Gopnik is one of the best scientist-authors around, translating complex study results into understandable ideas for parents. In this book, she encourages parents to view the parent-child relationship in a different way that what most parenting books would recommend. The book's title references the idea that parenting might be more about giving a child good soil for growth rather than a blueprint they need to follow.

For a deep dive into a child development topic:

Becoming Who We Are:
Temperament and Personality in Development
by Mary K. Rothbart

Widely recognized as a leading expert on temperament, Mary K. Rothbart has authored books, articles, and laboratory tools on this subject. She was also my grad school advisor and inspired me to learn about children in deep and detailed ways by observing how they respond to the world around them. This is a deep read, but you will understand temperament, your own child, and yourself in new ways by the end.

For Bibles that are age-appropriate and beautiful:

Frolic First Bible
by J. A. Reisch
and *Frolic Preschool Bible*
by Lucy Bell, illustrations by Natasha Rimmington

Full disclosure here, people: I helped develop these two Bibles! My team pored over every word and illustration to ensure that infants, toddlers, and preschoolers were introduced to God's Word through age-appropriate stories and pictures. Throughout this process, we thought about how this Bible would appeal to young children and to parents and caregivers like you who are guiding their children to take little steps toward big faith.